D1562600

MONOGRAPHS ON STATISTICS AND APPLIED PROBABILITY

General Editors

D.R. Cox, D.V. Hinkley, N. Reid, D.B. Rubin and B.W. Silverman

(Full details concerning this series are available from the Publishers.)

Transformation and Weighting in Regression

RAYMOND J. CARROLL
Texas A & M University

and

DAVID RUPPERT
Cornell University

CHAPMAN & HALL

I(T)P An International Thomson Publishing Company

New York • Albany • Bonn • Boston • Cincinnati
• Detroit • London • Madrid • Melbourne • Mexico City
• Pacific Grove • Paris • San Francisco • Singapore
• Tokyo • Toronto • Washington

Copyright © 1988

Chapman & Hall
A division of International Thomson Publishing Inc.
I(T)P The ITP logo is a trademark under license

Printed in the United States of America

For more information, contact:

Chapman & Hall
One Penn Plaza
New York, NY 10119

Chapman & Hall
2-6 Boundary Row
London SE1 8HN

International Thomson Publishing
Berkshire House 168-173
High Holborn
London WC1V 7AA
England

International Thomson Editores
Campos Eliseos 385, Piso 7
Col. Polanco
11560 Mexico D.F. Mexico

Thomas Nelson Australia
102 Dodds Street
South Merlbourne, 3205
Victoria, Australia

International Thomson Publishing Gmbh
Königwinterer Strasse 418
53228 Bonn
Germany

Nelson Canada
1120 Birchmount Road
Scarborough, Ontario
Canada, M1K 5G4

International Thomson Publishing Asia
221 Henderson Road
#05-10 Henderson Building
Singapore 0315

International Thomson Publishing-Japan
Hirakawacho-cho Kyowa Building, 3F
1-2-1 Hirakawacho-cho
Chiyoda-ku, 102 Tokyo
Japan

10 9 8 7 6 5 4 3 2

Library of Congress Cataloging-in-Publication Data

Carroll, Raymond J.
  Transformation and weighting in regression.
  (Monographs on statistics and applied probability)
  Bibliography: p.
  Includes index.
  1. Regression analysis. 2. Analysis of variance. 3. Estimation theory.
  I. Ruppert, David, 1948-   II. Title. III Series.
  ISBN 0-412-01421-1
  QA278.2.C37 1988
  519.5'36   87-23836
                                                         CIP

Please send your order for this or any Chapman & Hall book to **Chapman & Hall, 29 West 35th Street, New York, NY 10001, Attn: Customer Service Department.** You may also call our Order Department at 1-212-244-3336 or fax your purchase order to 1-800-248-4724.

For a complete listing of Chapman & Hall's titles, send your requests to **Chapman & Hall, Dept. BC, One Penn Plaza, New York, NY 10119.**

To our parents

Contents

Preface

This book is about the analysis of regression data when some of the classical assumptions are violated. Many of our consulting projects have involved regression modeling when the systematic part of the data had a fairly well understood functional form. Often the data have exhibited combinations of nonconstant variances, outliers, and skewness, and much of our work has been directed toward understanding how to fit models to such data. Our purpose in this monograph is to summarize the techniques we have found most useful for problems typically encountered.

The book contains three major parts. After an introductory chapter, Chapters 2 and 3 discuss the analysis of data when the primary concern is to overcome nonconstant variance in the responses. Chapter 2 describes estimation by weighted and generalized least squares, and provides an overview of the theory, graphical techniques, and applications. In Chapter 3, we discuss the estimation of variance functions. This chapter reviews a large and widely scattered literature.

The second part of the book discusses data transformations, particularly the technique we call 'transform both sides'. We have found this technique very useful when fitting nonlinear regression models to data that exhibit skewness and nonconstant variance. Chapter 4 is an introduction to data transformation in this context, while Chapter 5 combines weighting and data transformation.

The third part of the book concerns the detection of influential and outlying points and robust methods of estimation, i.e., methods that are not overly sensitive to outliers and other minor deviations from the ideal statistical model. At least five books have been written since 1980 on influence diagnostics or robust estimation in the linear regression model. We narrow our focus to just a few methods that we have found helpful when fitting transformation and variance models.

The book is a monograph, not a textbook. We hope that the material will be accessible to a wide range of people using regression, though some sections require a good statistical background. Each chapter introduces the material in a general and mostly nontechnical way, and includes at least one worked example along with details of estimation, inference, and plotting techniques. We have deliberately avoided a thorough mathematical treatment of the subject, not because such a development would be uninteresting but because we want our audience to be fairly wide. Some basic background on estimation and testing is sketched in Chapter 7.

We assume that the reader is familiar with linear and nonlinear regression, including the analysis of residuals. The reader should be acquainted with the material in a standard regression textbook, say Draper and Smith (1981) and Weisberg (1985), before beginning this book.

Nonlinear regression models in which the classical assumptions do not hold have been a major part of our research. D. Ruppert first encountered such data while a graduate student at the University of Vermont; a reanalysis of this 'bacterial clearance' data is given in Chapters 4, 5, and 6. Through the help of Rob Abbott and Kent Bailey, R.J. Carroll analyzed such data on a regular basis while at the National Heart, Lung, and Blood Institute, and an example of these data is given in Chapter 2. Together, we worked with Rick Deriso, of the International Pacific Halibut Commission, and Rod Reish, now at General Electric, on a long-term project modeling the Atlantic menhaden fishery, a project that led to our work on data transformations given in Chapter 4. We also acknowledge with gratitude the many examples brought to us by Wendell Smith, of Eli Lilly and Company, Perry Haaland, of Becton Dickenson Research Center, and Carl Walters, of the University of British Columbia. The bacterial clearance data were kindly supplied by George Jakab. Mats Rudemo gave us many insightful comments about his work with data transformations in agricultural and biological applications. We thank Brian Aldershof, Marie Davidian, Stena Kettl, Miguel Nakamura, Arny Stromberg, and Yin Yin for help in making the manuscript ready for publication.

R.J. Carroll's research was supported by the Air Force Office of Scientific Research, while D. Ruppert's was supported by the National Science Foundation.

North Carolina and Cornell   R.J. Carroll
1987   D. Ruppert

CHAPTER 1

Introduction

1.1 Preliminaries

When modeling data it is often assumed that, in the absence of randomness or error, one can predict a response y from a predictor x through the deterministic relationship

$$y = f(x, \beta) \tag{1.1}$$

where β is a regression parameter. The equation (1.1) is often a theoretical (biological or physical) model, but it may also be an empirical model that seems to work well in practice, e.g., a linear regression model. In either case, once we have determined β then the system will be completely specified.

Observed data almost never fit a model exactly, so that β cannot be determined exactly. Such a situation occurs when the relationship (1.1) holds exactly, but we are able to observe the response y or the predictor x only with error. This is the useful perspective behind the 'errors-in-variables' literature, although it is not the standard explanation why observed data do not fit models exactly. Most statistical analyses assume that we can measure the predictor x accurately, but that, given a value of x, the relationship with the observed response y is not deterministic. Partly this might be due to measurement error in y, but there are also other factors to consider. The lack of determinism could be a physical fact, e.g., the physical process generating an observed y from x might not be deterministic. Also, the model might not be exact either because of slight misspecification or because certain other predictors might have been omitted.

Understanding the variability of the response is important. Efficient estimation of the unknown parameter β requires knowledge of the structure of the errors. Also, understanding the distribution of y may be important in itself, particularly in prediction and calibration problems.

A good example of these issues is the relationship between the number of fish in a spawning population ('spawners' denoted by S) and the number of new fish eventually recruited into the fishery ('recruits' denoted by R). A standard theoretical deterministic model is due to Ricker (1954), and can be written as

$$R = \beta_1 S \exp(-\beta_2 S) = f_{RK}(S, \beta) \tag{1.2}$$

for $\beta_2 \geqslant 0$. This model is discussed at length in Chapter 4. In practice, the spawner–recruit relationship is not deterministic, and in fact there is considerable error in the equation. Partly, this is due to difficulties in measuring R and S, but perhaps even more important is the existence of unobserved environmental factors, e.g., ocean currents, pollution, predator depletion, and availability of food. Estimating the parameter β is important, but just as important for managing the fishery is understanding the distribution of the recruitment given the size of the spawning population. Only after this distribution is understood can one begin to assess the impact of different management strategies on the long-term productivity of the fishery.

In Ruppert *et al.* (1985) and Reish *et al.* (1985), a stochastic model for the Atlantic menhaden fishery is used to analyze the risk inherent in management strategies. The risk is largely due to the stochastic nature of recruitment, though estimation error in the parameters is also a factor. Certain harvesting policies, such as constant-catch policies where the same size catch is taken every year, are optimal for a deterministic fishery, but perform surprisingly poorly in stochastic situations. The main point is that the size and nature of management risk can only be assessed if a realistic model of recruitment variability (conditional on the size of the spawning population) is used.

1.2 The classical regression model

The usual regression model makes four basic assumptions and the analyses are often based on a fifth. The first four assumptions are that the model is correctly specified on average, and that the errors are independent, have the same distribution, and have constant variability, i.e.

$$E(y) = \text{expected value of } y = f(x, \beta) \tag{1.3}$$

$$y - f(x, \beta) = \varepsilon, \text{ Variance}(y) = \text{Variance}(\varepsilon) = \sigma^2 \tag{1.4}$$

The errors ε have the same distribution for each
given value of x (1.5)

Given x, the errors $\varepsilon = y - f(x, \beta)$ are independently
distributed   (1.6)

Of course, (1.5) implies (1.4). Sometimes only (1.4) is assumed, so we list it separately. Most often, the parameter β is estimated by least squares, the basic motivation for which is that it is efficient if the errors are normally distributed. This leads to the fifth assumption, i.e.

The errors are normally distributed   (1.7)

To see how these assumptions might work in practice, we consider the spawner–recruit relationship for the Skeena River sockeye salmon. The data are described in Chapter 4 and are plotted in Figure 4.1. In Figure 4.3, we plot the predicted values against the residuals. An examination of these two figures suggests that at least two of the major classical assumptions are strongly violated. The first assumption (1.3) seems reasonable, as the Ricker curve appears adequate. Whether the independence assumption (1.6) holds is not clear here, but it can be taken as a working hypothesis especially in light of the small size of the data set. A check of the serial correlations of the residuals suggests that (1.6) is not seriously violated. As can be seen from either figure, the variability of the data increases as a function of the size of the spawning population, thus violating assumptions (1.4) and (1.5). The normality assumption (1.7) seems adequate, although there may be some right-skewness.

The major purpose of this book is to discuss methods for analyzing regression data when the classical assumptions of constant variance and/or normality are violated.

Having observed that the variances are not constant, we might take as a working hypothesis that the variances are proportional to the square of the mean. This is a model with a constant coefficient of variation (standard deviation divided by mean). This model involves no additional parameters, but requires the a priori assumption about the variance function. A more flexible approach is to use a variance model with unknown parameters, for example to assume that the variance is proportional to some (unknown) power of the mean. To obtain a better estimate of the regression parameter β and to understand the distribution of the size of the recruit population given a spawning population size, two basic strategies can be employed.

First, we might continue to assume normally distributed observations. If the variance function contains unknown parameters, these must be estimated, perhaps by using residuals from an unweighted fit. After obtaining a fully specified variance model, some form of iterative weighted least squares can be employed to estimate the regression parameter β.

The second basic strategy is to replace the normality assumption by assuming a distribution family that has a constant coefficient of variation. Even within this strategy there are at least two widely used options. The first is to assume that the responses are gamma distributed. Since the gamma family is an exponential family of distributions, this puts us into the class of generalized (non)linear models. These have been admirably discussed by McCullagh and Nelder (1983). Luckily, the fitting method for estimating β will be the same iterative weighted least-squares algorithm as in the case of normally distributed observations. The second common option is to assume that the data have a lognormal distribution, i.e., the logarithm of the response is normally distributed. This is an example of a transformation model, where we assume that

$$\ln(R) = \ln[f_{RK}(S, \beta)] + \varepsilon \qquad (1.8)$$

where the errors ε satisfy the classical assumptions, at least approximately. In other words, a transformation model says that we can transform the data and the model in such a way as to return to the classical assumptions.

1.3 The scope of the book and an outline

This book examines the analysis of data in which the classical regression assumptions of constant variance and normally distributed data are violated. We hope to provide a fairly comprehensive review of the methods we have found useful in analyzing such data. These methods fit naturally into two broad categories: methods based on weighting and weighted least squares, and methods based on data transformations.

Standard methods of weighting and data transformation can be adversely affected by just a few observations, especially by outliers. In fact, the situation here is much worse than when fitting ordinary regression models. The influence functions of variance-parameter estimators are quadratic in the response rather than linear as in the case of regression parameters. Generally, the influence function of a transformation-parameter estimator also increases at a rate faster than linear. Therefore weighting and transformation methods should be accompanied by diagnostics to identify influential observations and by robust methods to fit the data in the presence of such observations. Outlying data should not be detected simply to be downweighted or removed. Outliers can indicate model inadequacies and may suggest useful improvements to the model. However, few statisticians would increase the complexity of a model to fit an outlier unless the new model also gave a better fit to the bulk of the data. Model selection and the identification of outliers are intrinsically connected. We make this point clear in the examples of Chapter 6.

1.3.1 Methods based on weighting

One method for analyzing data with nonconstant or heteroscedastic variances is based on iterative weighted least squares after hypothesizing and fitting a model for the variances. The constant-coefficient-of-variation model discussed in relation to the Skeena River data states that

$$\text{Variance}(R) = \sigma^2 f_{RK}^2(S, \beta)$$

This is not the only variance model that we might consider. A simple generalization is to suppose that the variance is proportional to a power of the mean response

$$\text{Variance}(R) = \sigma^2 f_{RK}^{2\theta}(S, \beta) \tag{1.9}$$

In this case, $\theta = 1$ would be appropriate for gamma or lognormal data, while $\theta = 0.5$ would arise if recruitment had a Poisson distribution. It might happen that neither $\theta = 0.5$ nor $\theta = 1$ is appropriate, and we might let the data give us some indication of the value of θ. In radioimmunoassay experiments, where the response is a count from a binding experiment and the predictor is the concentration of a substance, it is often found that model (1.9) holds but with θ between 0.6 and 0.8. Alternatively, one might let the variance be a more direct function of the predictor, as in

$$\text{Variance}(R) = \sigma^2(1 + \theta_1 S + \theta_2 S^2)$$

All of these variance models can be summarized as belonging to the class

$$\text{Variance}(R) = \sigma^2 g^2(S, \beta, \theta) \qquad \text{for some function } g$$

In Chapter 2 we discuss the problem of estimating the regression parameter β when the variance parameter θ is known. We spend some time discussing generalized least squares, the most common method of estimating β. This method is used not only when normality of the responses is assumed, but also if one is in the class of generalized linear models. Our focus is on the former case because the recent book by McCullagh and Nelder (1983) covers the latter thoroughly. Generalized least squares has the advantage that it can be applied without any distributional assumptions; one need only specify the model for the mean and the variances. We also discuss methods of inference and calibration. Section 2.7 describes a few graphical techniques for detecting and understanding heteroscedastic variability. Section 2.8 contains worked examples, while in section 2.9 we show that modeling the variances adequately can have beneficial effects when prediction intervals are required.

Usually, the variance parameter θ is not known in practice and must be estimated. There are many ways of doing this. Chapter 3 reviews the most common methods for estimating variance functions in regression, along with hypotheses tests and confidence intervals for the variance parameter. The emphasis is on unifying a widely scattered literature, and we especially make use of the fact that most such methods of estimation are closely related to regression, but where now the 'responses' are functions of residuals from a preliminary fit. The examples of Chapter 2 are reinvestigated by applying some of the methods of estimation to data.

1.3.2 Methods based on data transformation

The generalized least-squares methods of Chapters 2 and 3 attack the problem of variance heterogeneity directly by specifying models for both the mean and variance, and then estimating the relevant parameters simultaneously. The alternative approach commonly used in applied statistics is to transform the data in such a way as to obtain a new regression model with clearly defined properties, e.g., constant variance and/or normally or symmetrically distributed errors. Throughout this book, we are working in the context that one already has a model relating the predictors to the response in the absence of any error. This model could be determined from physical principles or convention, but the key point is that it describes the systematic structure of the data. Of course, having a model does not excuse one from checking its correctness, but when the model is treated as a working hypothesis we certainly do not want a data transformation methodology that invalidates the model. Thus, when we used the logarithmic transformation in (1.8), what we did was to transform both sides of the multiplicative model

$$R = f_{RK}(x, \beta)\eta \qquad (1.10)$$

where η is lognormally distributed. This is an example of the transform-both-sides approach to nonlinear regression models (see Carroll and Ruppert, 1984a). Since the Ricker model has some biological plausibility, it makes little sense to transform the left-hand side of (1.10) without using the same transformation on the right-hand side.

Transformations other than the logarithm can be used to induce constant variance or approximate normality, including the square root or the inverse. All of these transformations along with the option of not transforming at all can be gathered up into a single model

$$h(R, \lambda) = h(f_{RK}(S, \beta), \lambda) + \varepsilon$$

where the function $h(v, \lambda)$ is the Box and Cox (1964) power transformation family

$$h(v, \lambda) = \begin{cases} (v^\lambda - 1)/\lambda & \text{if } \lambda \neq 0 \\ \ln(v) & \text{if } \lambda = 0 \end{cases}$$

Choosing $\lambda = 0$ would mean taking logarithms, $\lambda = 0.5$ would be square roots, and $\lambda = 1$ would mean no transformation at all.

In Chapter 4 we explore the transform-both-sides model in detail. Methods of estimating the transformation are discussed, along with methods for making inference about the parameter β. The Skeena River data are reanalyzed without the restriction that a logarithmic transformation be used. We also describe analyses of two other data sets, one a spawner–recruit study on another fishery and the second a study of bacterial clearance in the lungs of mice. In the latter case the transformation removes the skewness and some of the heteroscedasticity, but after transformation the variance still depends strongly on a covariate, time to sacrifice. This residual heteroscedasticity motivates Chapter 5.

In Chapter 5 we combine the ideas of weighting and transformation. Since both approaches are trying to solve the same problem it is useful to consider how to unify them. The Skeena River and bacterial clearance examples are reanalyzed from this perspective, and especially in the latter we find that the combination yields a useful analysis.

1.3.3 Regression diagnostics and robust estimation

The methods of estimation and inference in common use in regression are sensitive to minor problems with the data. These problems might include recording errors as well as natural phenomena, so that the deletion or modification of only a few or even a single observation can change the whole pattern of analysis or even basic conclusions. In the constant-variance regression model, there are a variety of tools that one can use to protect oneself against being misled by a few data points. The most widely implemented are methods of regression diagnostics, as represented by the recent books of Belsley et al. (1980), Cook and Weisberg (1982), and Atkinson (1985). The aim here is to find those points whose modification or deletion would have a major impact on the analysis. Less widely implemented but still valuable are the methods of robust estimation, recent books including those by Huber (1981) and Hampel et al. (1986). Robust estimation methods attempt to construct estimates of parameters which are not sensitive to the deletion or modification of a few points. It often happens that regression diagnostics started from robust rather than the usual methods more readily identify unusual observations.

The weighting and transformation methods in common use are decidedly not robust, sometimes in unusual ways. For example, in weighted regression it is our experience that the most troublesome observations are those with the smallest predicted variance. Influence diagnostics and robust estimates are required. This is the content of Chapter 6. Rather than trying to replicate the five published books on diagnostics and regression, we decided to present a single diagnostic and a single estimation scheme, both of which we have found useful in practice. These are applied to the bacterial clearance data and lead to a useful modification of the variance model.

1.3.4 Some notes on computation

The computations were done on IBM PC/AT personal computers using either the GAUSS matrix programming language or APL through the STATGRAPHICS package, the latter mostly for plots and the nonparametric regression estimators we used in our analyses.

CHAPTER 2

Generalized least squares and the analysis of heteroscedasticity

2.1 Introduction

Fitting a nonlinear regression model is often complicated by heterogeneous rather than constant variability. A particular example of this phenomenon is the analysis of assay data (see Finney, 1976; Rodbard, 1978; Raab, 1981a, b; Myers *et al.*, 1981; Butt, 1984). In assays, the concentration or dose is considered to be the predictor and the response is often a count, and with count data we can expect that the variability depends on the mean. Heterogeneous variability, often called heteroscedasticity, arises in almost all fields. We have seen examples in: (1) pharmacokinetic modeling (Bates *et al.*, 1985), where the variability depends on time; (2) analysis of the stability of drugs (Kirkwood, 1977, 1984; Nelson, 1983; Tydeman and Kirkwood, 1984), where the variability depends on time and storage temperature; (3) enzyme kinetics (Cressie and Keightley, 1979, 1981; Currie, 1982), where the variability depends on concentration; (4) chemical reaction kinetics (Box and Hill, 1974; Pritchard *et al.*, 1977); (5) econometrics, with variability depending on any number of factors; (6) screening designs in quality control (Box and Meyer, 1985b; Box, 1987; Leon *et al.*, 1987); (7) fisheries research, where the variability in the production of new fish depends on the size of the spawning population (Ruppert and Carroll, 1985).

The general heteroscedastic nonlinear regression model for a response vector $Y = (y_1, \ldots, y_N)^T$ has two major components, the mean vector and the covariance matrix. The mean vector $\mu = (\mu_1, \ldots, \mu_N)^T$ has individual elements $\mu_i = f(x_i, \beta)$, where by convention f is called the regression function, β is the regression parameter and is a vector with p elements, and the predictors x_i^T when set out as row vectors are stacked into a matrix X with N rows. The mean values of the responses are determined by the regression parameter β. The covariance matrix for Y is the $N \times N$ matrix $\sigma^2 \Lambda$, where we write $\Lambda = \Lambda(\mu, Z, \theta)$ to emphasize the possible dependence of the covariance matrix on the mean vector μ, the structural variance parameter θ and a matrix of known variables $Z = (z_1, \ldots, z_N)$ whose individual component vectors z_i may or may not include some or all of the predictors x_i.

The key feature to modeling the means is the assumption that the mean vector varies systematically and smoothly as we perturb predictors. This assumption implies in particular that the regression function should do more than simply 'connect the dots' between neighboring observed responses. In simple straight-line regression the scatterplot of the predictors against the responses helps us determine an appropriate model. In many applications, instead of purely empirical models there are natural models that are suggested by the basic science of the problem, for example modeling enzyme reactions by the Michaelis–Menten equation

$$Y = \alpha_0 x / (\alpha_1 + x)$$

(see Cressie and Keightley, 1981; Currie, 1982).

Modeling the variability in data has received less attention. We view heteroscedasticity of variance as a regression problem, i.e., systematic and smooth change of variability as predictors are perturbed. Looked at in this way, there are many similarities with modeling the mean vector. In particular, the view that variability changes systematically and often smoothly with predictors dictates the need to go beyond the 'connect-the-dots' strategy one usually sees employed. The residual plots to be discussed in section 2.7 will replace the ordinary scatterplot as devices for detecting heterogeneity of variance. By analogy with regression for means, we will use the residual plots to suggest simple models for the variability, and we can suggest an analog for the straight-line fit. In Chapter 3, we discuss how one might fit the parameters in the variance relationship.

For the most part, we consider the case that the responses $y_1, \ldots, y_N$ are independently distributed, making the covariance matrix diagonal. Some discussion of the general case is given in section 2.6, but readers looking for a definitive treatment on such cases as variance components will not find it here. Throughout this chapter, we will write the model as

$$E(y_i) = \mu_i(\beta) = f(x_i, \beta)$$
$$\text{Variance}(y_i) = \sigma^2 g^2(\mu_i(\beta), z_i, \theta) = \sigma^2 / w_i \qquad (2.1)$$

The constants w_i are commonly called the true weights, a terminology we explain below.

When modeling the variance, it is useful to distinguish four major cases based on (a) whether the structural variance parameter θ is known or not and (b) whether the variance function g does or does not depend on the mean values μ_i. The easiest case to handle occurs when the weights $(w_1, \ldots, w_N)$ in (2.1) are known, i.e., θ is known and the variability is not a function of the mean. Here is a simple trick that returns us to the ordinary nonlinear regression model. Redefine the model

$$y_i^* = w_i^{1/2} y_i$$
$$f^*(w_i, x_i, \beta) = w_i^{1/2} f(x_i, \beta) \qquad (2.2)$$

The redefined responses y_i^* now have constant variances σ^2 with means given by the new nonlinear function f^*. The usual method of estimation is least squares, thoroughly discussed in a modern treatment by Bates and Watts (1988). Here the idea is to minimize the sum of the squared deviations, i.e.

$$\text{Minimize in } \beta \text{ the expression } \sum_{i=1}^{N} [y_i^* - f^*(w_i, x_i, \beta)]^2$$

When we translate this back to the original model, we see that our estimation problem is

$$\text{Minimize in } \beta \text{ the expression } \sum_{i=1}^{N} w_i [y_i - f(x_i, \beta)]^2 \qquad (2.3)$$

The interpretation of (2.3) as weighted least squares is natural: the larger the values of the weight w_i, the larger the contribution of that squared deviation. When we note that the weights are the inverses of the variances, we see that weighted least squares gives most weight to data points with low variability, which seems desirable.

When the model (2.1) holds, for sufficiently large sample sizes the weighted least-squares estimate of β will be less variable than the unweighted least-squares estimate; this is just the Gauss–Markov theorem and a linearization argument. For large enough sample sizes the weighted least-squares estimate will be asymptotically normally distributed with mean β and covariance matrix $(\sigma^2/N)S_G^{-1}$, where

$$S_G = N^{-1} \sum_{i=1}^{N} f_\beta(x_i, \beta) f_\beta(x_i, \beta)^T / g^2(\mu_i(\beta), z_i, \theta) \qquad (2.4)$$

and f_β is the derivative of f with respect to β.

In practice the variances and hence the correct true weights w_i are rarely known *a priori*. Since estimating the structural parameter θ presents its own special problems, we begin our discussion with the case that the variance depends on the mean but θ is known. Besides ease of presentation, assuming that θ is known allows us to consider many issues of estimating β in the presence of heterogeneous variance without the need to refer to an algorithm for estimating θ. The most common model for variability as a function of the mean assumes that the standard deviation is proportional to a power θ of the mean response

$$\text{Standard deviation } (y_i) = \sigma\mu_i(\beta)^\theta \tag{2.5}$$

The Poisson model follows (2.5) with $\theta = 0.5$, while the gamma and lognormal distributions have $\theta = 1.00$. The model (2.5) with θ unknown is in common use in radioligand assays and kinetic reaction experiments, among other fields (see Box and Hill, 1974; Finney, 1976; Pritchard *et al.*, 1977; Raab, 1981a).

It is sometimes suggested that one estimate variances and weights based on replication, taking m replicate responses at each design point x_i and estimating the weights by the inverse of the sample variances. This method is unsatisfactory because it tells us nothing about the structure of the variances. Worse, unless the number of replicates m is large, this can be a very inefficient practice. The trouble is that sample variances based on small degrees of freedom are wildly unstable, so that one is introducing unnecessary variability into the problem. As in Bartlett (1947), Neyman and Scott (1948), and Fuller and Rao (1978), the fully nonparametric use of replicates to estimate weights results in estimates of the regression parameter β that are more variable than weighted least squares with known weights. Of course, the extra variability is largest when the number of replicates is small. The problem with taking a large number of replicates is that one then can investigate only a small number of design points, since variances stabilize only when using a fairly large number of replicates (e.g., 10) per point. A common theme of this and the next chapter is that one need not take so many replicates, provided one assumes some underlying smoothness of the variance function. By viewing the estimation of variances as a form of regression, we can analyze dispersion without insisting on replicates.

2.2   Estimation by generalized least squares

When the weights w_i in (2.1) are unknown, the most common method of estimating parameters in the mean function is to apply weighted least squares with the estimated weights $\hat{w}_i$. This method is called *generalized least squares*. Generalized least squares makes no assumptions about the underlying distributions for the data, and instead relies only on the mean and variance relationship (2.1). This can be looked at as either an advantage or a disadvantage. The major advantage is the semiparametric feature; generalized least squares enables us to understand the systematic aspects of the data without becoming involved in higher-level modeling, the distributions generally being harder to identify.

Of course, if one does have knowledge of the underlying distribution, then one can in principle use this knowledge to obtain even more efficient estimates of β and θ. The last argument is seductive but tricky. As we shall discuss in section 2.4, for models such as the Poisson and the gamma distribution, generalized least-squares estimates of β are asymptotically equivalent to the efficient maximum-likelihood estimates. In other instances, maximum likelihood is more efficient than generalized least squares when the assumed distribution is correct, but otherwise can be disadvantageous (see section 2.4).

There are as many generalized least-squares estimates as there are schemes to generate estimated weights. For example, one might start off with the unweighted least-squares estimate of β, denoted by $\hat{\beta}_L$. One might then estimate the weights as

$$\hat{w}_i = 1.0/g^2(\mu_i(\hat{\beta}_L), z_i, \theta)$$

forming a generalized least-squares estimate $\hat{\beta}_G$. Intuitively, the new estimate should be more efficient than unweighted least squares, so we might form new weights starting from $\hat{\beta}_G$ and possibly even consider iterating the process. We formalize these ideas in the following algorithm.

Algorithm for generalized least squares

Step 1   Start with a preliminary estimator $\hat{\beta}_*$.

Step 2   Form the estimated weights

$$\hat{w}_i = 1.0/g^2(\mu_i(\hat{\beta}_*), z_i, \theta) \tag{2.6}$$

Step 3   Let $\hat{\beta}_G$ be the weighted least-squares estimate using the estimated weights (2.6).

Step 4   Update the preliminary estimator by setting $\hat{\beta}_* = \hat{\beta}_G$, and update the weights as in (2.6).

Step 5   Repeat steps 3 and 4 $\mathbb{C} - 1$ more times, where $\mathbb{C}$ is the number of cycles in generalized least squares and is chosen by the experimenter.

Step 6   Full iteration corresponds to setting the number of cycles $\mathbb{C} = \infty$. This is the same as solving the equations

$$0 = \sum_{i=1}^{N} f_\beta(x_i, \beta)[y_i - f(x_i, \beta)]/g^2(\mu_i(\beta), z_i, \theta) \qquad (2.7)$$

which will usually be done by a separate numerical method. *End*

Fully iterating steps 1 to 5 is a version of iteratively reweighted least squares. Faster computation not requiring repeated nonlinear least-squares minimizations are possible using the Gauss–Newton form of iteratively reweighted least squares as in McCullagh and Nelder (1983, p. 172). Although iterative weighted least squares need not converge, in our experience convergence is usual.

What should the starting estimate in step 1 be, and how many cycles $\mathbb{C}$ are necessary in the generalized least-squares algorithm? The obvious approach toward answering this question is to contrast the distributions of the resulting estimators. Finite-sample theoretical calculations are difficult, and the natural step is to investigate asymptotic distributions. Unfortunately, standard asymptotic theory is noninformative since all generalized least-squares estimates have the same asymptotic distribution. This fact is a folklore theorem, well known for many years by econometricians and proved precisely by Jobson and Fuller (1980) and Carroll and Ruppert (1982a, b), with informal statements and proofs given by Carroll (1982a), McCullagh (1983), and the argument in section 7.1.

Theorem 2.1   The basic theorem of generalized least squares
For any $N^{1/2}$-consistent starting estimate $\hat{\beta}_*$ and any number of cycles $\mathbb{C}$, the generalized least squares estimate is asymptotically normally distributed with mean β and covariance matrix $(\sigma^2/N)S_G^{-1}$, where S_G is given by (2.4).

Theorem 2.1 explains why there has been no clear consensus about the best choice of the number of cycles $\mathbb{C}$. Goldberger (1964) and Matloff *et al.* (1984) propose $\mathbb{C} = 1$ cycle, the latter stating that 'the first iteration is usually the best, with accuracy actually deteriorating as the number of cycles is increased'. Williams (1959, pp. 20 and 70) and Seber (1977) clearly suggest that the larger the number of iterations the better, but they finally recommend $\mathbb{C} = 2$, with Williams stating that 'only two cycles of iterations are usually required ... since great accuracy in the weights is not necessary ...'. Wedderburn (1974), McCullagh (1983), and McCullagh and Nelder (1983) only consider the possibility of fully iterating; setting $\mathbb{C} = \infty$ is a special case of generalized least squares which is equivalent to quasi-likelihood estimation (see section 2.5). In the next section we explore the choice of the number of cycles in detail.

2.3 Simulation studies of generalized least squares

If $\hat{\beta}_G(\mathbb{C})$ is a generalized least-squares estimate based on $\mathbb{C}$ cycles of the algorithm, the effect of the number of cycles as the sample size becomes large shows up in the following way. The estimate has covariance matrix given as

$$\text{Covariance matrix of } N^{1/2}[\hat{\beta}_G(\mathbb{C}) - \beta] = \sigma^2 S_G^{-1} + N^{-1} V_N(\mathbb{C})$$
$$+ \text{(smaller-order terms)}$$
$$(2.8)$$

The effect of the number of cycles in the algorithm shows up in the correction terms $V_N(\mathbb{C})$, and at least in principle we can distinguish between different values of $\mathbb{C}$ by comparing the correction terms.

For the case of linear regression, Carroll *et al.* (1987) have shown that the correction term $V_N(1)$ for one cycle differs from the correction terms for two or more cycles, the latter of which are equal. Thus, Williams' statement that two cycles suffice has some basis. Unfortunately, it is not always true that the correction term for one cycle is bigger than the correction terms for two or more cycles; the indeterminate nature of the theory carries over into simulation experiments.

The major simulation evidence concerning this question is the paper by Matloff *et al.* (1984). They study the power-of-the-mean model (2.5) for normally distributed data. For $\sigma = 1.0$ and $\theta = 0.5$ their results suggest that the number of cycles is not important, while for $\sigma = 1.0$ and $\theta = 1.0$ there is a strong dependence of the results on the choice of the number of cycles, with the choice $\mathbb{C} = 1$ being clearly the best.

We are not certain what causes these differences. On a heuristic level, there are two components of the data which determine the behavior of generalized least squares in power-of-the-mean models. First, of course, is the relative spread in the standard deviations as we move through the data. A reasonable rule of thumb is that, if the standard deviations do not vary by a factor of 1.5:1 in the data, then weighting is generally unlikely to be of much importance, while if the standard deviations vary by a factor of 3:1 or more, then weighting will generally be called for. A second factor of importance is the size of the standard deviations relative to the means, which we like to express as the ratio

$$\frac{\text{Median estimated standard deviation}}{\text{Median absolute estimated mean}} = \frac{\text{Median } \{\hat{\sigma}g(x_i, \hat{\beta}, \theta)\}}{\text{Median } |f(x_i, \hat{\beta})|}$$

When this median coefficient of variation is small, the number of cycles of generalized least squares tends to be less important. In our experience, the median coefficient of variation is often less than 0.35. This is not always the case, of course. For example, the underlying observations that lead to the car insurance data listed in McCullagh and Nelder (1983, Chap. 6) appear to have coefficient of variation nearly 1.0. This measure of relative variation can be manipulated by adding a constant to the regression function (although then the model changes!), so one must take care not to rely on it too heavily. On an overall basis, we have found it to be rather useful in the power-of-the-mean model.

It is instructive to re-examine the simulation study of Matloff et al. (1984) and compute their relative spread of the standard deviations and median coefficients of variation. For one part of their study they examine the linear model with means

$$\beta_0 + \beta_1 x_i^{(1)} + \beta_2 x_i^{(2)}$$

where $\beta_0 = 1.0$, $\beta_1 = \beta_2 = 10.0$, and the predictors $\{x_i^{(1)}, x_i^{(2)}\}$ were independently normally distributed with mean 5.0 and standard deviation 2.0. When $\theta = 0.5$ and $\sigma = 1.0$, the ratio of the 90th percentile to the 10th percentile of the standard deviations is 1.46, and the median coefficient of variation is 0.07. Based on our previous discussion, we should not expect to see much improvement by weighting and we should not expect to see much effect due to the number of cycles in the generalized least-squares algorithm; both expectations are met in the simulation. When $\theta = 1.0$ and $\sigma = 1.0$, the median coefficient of variation is 1.0 and the ratio of the 90th percentile to the 10th percentile of the standard deviations is 2.13. Thus, we would expect to see some effect due to weighting and the choice of the number of cycles, and our expectations are not disappointed.

We have redone this part of the simulations but with much smaller coefficients of variation. We continue with the model (2.5) with $\theta = 1.0$, but now we take $\sigma = 0.10, 0.20$, and 0.30. The sample size we took was $N = 20$, and there were 500 simulations in the experiment. The results are reported in Table 2.1. We also repeated the study with a regression based on the design and parameters given in the paper by Jobson and Fuller (1980); see Table 2.2.

It is apparent from Tables 2.1 and 2.2 that the situation studied by Matloff et al. (1984) differs from problems with smaller coefficient of variation. While they find increased mean squared errors of more than 250% caused by having to estimate the variance function, we find

Table 2.1   *A modification of a simulation study performed by Matloff et al. (1984). Their results for coefficient of variation $CV = 1.00$ are listed under MRT*

Coefficient of variation, CV	Cycles in generalized least squares, $\mathbb{C}$	Mean squared errors as a fraction of the exact MSE with known weights	
		β_0	Average β_1, β_2
0.10	1	1.01	1.00
0.10	2	1.01	1.01
0.10	5	1.01	1.01
0.20	1	1.05	1.02
0.20	2	1.06	1.05
0.20	5	1.06	1.06
0.30	1	1.13	1.08
0.30	2	1.17	1.15
0.30	5	1.20	1.20
MRT	1	3.72	3.91
MRT	2	5.50	3.90
MRT	5	5.26	4.46

Table 2.2   *A modification of a simulation study performed by Matloff et al. (1984). The design was as in Jobson and Fuller (1980), as were the regression parameters*

Coefficient of variation, CV	Cycles in generalized least squares, $\mathbb{C}$	Mean squared errors as a fraction of the exact MSE with known weights		
		β_0	β_1	β_2
0.10	1	1.03	1.03	1.03
0.10	2	1.01	1.01	1.01
0.10	5	1.01	1.01	1.01
0.20	1	1.09	1.09	1.09
0.20	2	1.04	1.04	1.05
0.20	5	1.04	1.03	1.05
0.30	1	1.19	1.18	1.17
0.30	2	1.10	1.09	1.12
0.30	5	1.09	1.08	1.11

increases of the order 10%. We especially do not find very large differences caused by choosing a different number of cycles $\mathbb{C}$ in the generalized least-squares algorithm. It is interesting to note that, in Table 2.1, there is a tendency for increasing mean squared errors with increasing $\mathbb{C}$, with just the opposite in Table 2.2. To us, this merely reflects the indeterminate nature of the theoretical calculations.

To summarize the discussion, the second-order theory and the simulations agree qualitatively on a number of points. Doing only one cycle in generalized least squares differs from two or more cycles, but the direction of the difference is not always in favor of the latter. The larger the coefficient of variation, the larger the costs due to estimating the variance function and the larger the difference between using $\mathbb{C} = 1$ and $\mathbb{C} \geqslant 2$ cycles. We recommend at least two cycles of generalized least squares, largely to eliminate the effect of the inefficient unweighted least-squares estimate.

2.4   Maximum likelihood as an alternative to generalized least squares

The power of generalized least squares is that it enables us to understand the systematic features of the data without the need to make any further assumptions. Further, the estimates are easily computed and have a workable large-sample theory which enables one to make inferences about the regression parameter β (see section 2.5).

There are many cases where it will be reasonable to assume an approximate distribution for the data and then estimate parameters via maximum likelihood. For example, the mean and the variance of the Poisson distribution both equal λ, so that model (2.5) holds with $\theta = 0.5$ and $\sigma = 1$. If the mean λ is large, Poisson data will be approximately normally distributed. The loglikelihood for the data except for a constant not depending on β is

$$L(\beta) = \sum_{i=1}^{N} \{ y_i \log[f(x_i, \beta)] - f(x_i, \beta) \} \qquad (2.9)$$

Taking derivatives with respect to β and solving for zero, we see that the maximum-likelihood estimate of β satisfies

$$\sum_{i=1}^{N} f_\beta(x_i, \beta)[y_i - f(x_i, \beta)]/f(x_i, \beta) = 0 \qquad (2.10)$$

This is exactly a form of equation (2.7), so that maximum-likelihood estimation of β in the Poisson case is a generalized least-squares estimate based on full iteration; this fact is well known (see McCullagh and Nelder, 1983).

A second example is the gamma distribution, which has constant coefficient of variation σ and the probability density function

$$[y\Gamma(v)]^{-1}(vy/\mu)^v \exp(-vy/\mu) \qquad \sigma^2 = 1/v \qquad y \geqslant 0$$

The mean of the gamma distribution is μ, and its standard deviation is $\sigma\mu$. The skewness is 2σ, while the kurtosis is $6\sigma^2$. Thus the smaller the value of σ, the more closely the gamma density approximates a normal density. For fixed σ, the maximum-likelihood estimate of β satisfies

$$\sum_{i=1}^{N} f_\beta(x_i, \beta)[y_i - f(x_i, \beta)]/f^2(x_i, \beta) = 0$$

which is a case of (2.7). Again, the maximum-likelihood estimate here is a generalized least-squares estimate based on full iteration.

These two examples indicate that generalized least-squares estimates might be maximum-likelihood estimates for a wide class of models. The Poisson and gamma distributions are examples of distributions in the exponential family (see McCullagh and Nelder,

1983). The probability density or mass function of these distributions is given as

$$\exp\{[y\eta - b(\eta)]/\sigma^2 + c(y, \sigma)\} \qquad (2.11)$$

for some functions $b(\cdot)$ and $c(\cdot)$. The natural parameter η is related to the regression function $f(x, \beta)$ by

$$E(y) = f(x, \beta) = \frac{\partial}{\partial \eta} b(\eta)$$

If $f(x, \beta) = f(x^T \beta)$, then we have a generalized linear model. More generally, we will call this a generalized nonlinear model. It is well known that the variances of the responses are given by

$$\text{Variance}(y) = \sigma^2 \frac{\partial^2}{\partial \eta^2} b(\eta)$$

Write the second derivative with respect to η of $b(\eta)$ in the more general form

$$\frac{\partial^2}{\partial \eta^2} b(\eta) = g(\mu(\beta), z, \theta) \qquad (2.12)$$

Thus, the mean and variance function of a generalized nonlinear model form a special case of our general mean and variance model (2.1). For Poisson data, $\sigma = 1$, $\eta = \log\{f(x, \beta)\}$ and $b(\eta) = \exp(\eta)$. For gamma data $\eta = -1/f(x, \beta)$ and $b(\eta) = -\log(-\eta)$. For generalized nonlinear models, the log-likelihood of an observation y is

$$l(y, \beta, \sigma) = [y\eta - b(\eta)]/\sigma^2 + c(y, \sigma)$$

Since

$$\frac{\partial}{\partial \eta} f(x, \beta) = \frac{\partial^2}{\partial \eta^2} b(\eta) = \frac{\text{Variance}(y)}{\sigma^2} = g^2(\mu(\beta), z, \theta)$$

we have that

$$\frac{\partial}{\partial \beta} l(y, \beta, \sigma) = \frac{[y - f(x, \beta)]f_\beta(x, \beta)}{g^2(\mu(\beta), z, \theta)}$$

The maximum-likelihood estimate of β solves the equation

$$0 = \sum_{i=1}^{N} \frac{\partial}{\partial \beta} l(y_i, \beta, \sigma)$$

$$= \sum_{i=1}^{N} [y_i - f(x_i, \beta)] f_\beta(x_i, \beta)/g^2(\mu_i(\beta), z_i, \theta)$$

This is exactly equation (2.7), so that generalized least squares is equivalent to maximum-likelihood estimation in the class of generalized nonlinear models.

Firth (1987) shows that quasi-likelihood estimates are maximum-likelihood estimates for a natural exponential family, but otherwise the former are less efficient.

Neither a normal model with constant coefficient of variation nor a lognormal model is in the class of generalized nonlinear models. Both of these models have constant coefficient of variation, and hence would be competitors with the gamma model. For coefficient of variation σ smaller than about 0.3, Atkinson (1982) and Kotz and Johnson (1985) indicate that these models will be essentially indistinguishable from one another. Even for larger values of σ, where skewness begins to become pronounced, it may be difficult to discriminate between gamma and lognormal densities.

A method for estimating the regression parameter which serves as a competitor to generalized least squares is to assume that the data are normally distributed, follow the model (2.1), and then use maximum likelihood. This has been suggested as a general technique by Beal and Sheiner (1985) and Stirling (1985). Since there is information about the regression parameter β in both the mean and the variances, it is reasonable to conjecture that normal-theory maximum likelihood gives more efficient estimators. While this is the case for normally distributed data, the benefits of using maximum likelihood are bought by incurring losses if the model does not hold exactly.

The following discussion is based on the work of McCullagh (1983) for the one-sample problem and of Jobson and Fuller (1980) and Carroll and Ruppert (1982b) for regression (see also Amemiya, 1973). Suppose that we have independent and identically distributed data which we believe to be normally distributed with mean μ and standard deviation $\sigma\mu$, where the coefficient of variation σ is known. The generalized least-squares estimate is the sample mean $\bar{y}$, which is normally distributed with mean μ and variance $\sigma^2\mu^2/N$. The maximum-likelihood estimate is given by

$$\hat{\mu}_{\text{MLE}} = [(\bar{y}^2 + 4\sigma^2 T)^{1/2} - \bar{y}]/(2\sigma^2)$$

where

$$T = N^{-1} \sum_{i=1}^{N} y_i^2$$

For large sample sizes the maximum-likelihood estimate is normally distributed with mean μ and the ratio of the variances is approximately

$$\frac{\text{Variance}(\hat{\mu}_{\text{MLE}})}{\text{Variance}(\hat{\mu}_{\text{GLS}})} \simeq \frac{1 + \sigma^2(2 + \gamma_2) + 2\sigma\gamma_1}{(1 + 2\sigma^2)^2} \qquad (2.13)$$

where γ_1 and γ_2 are the standardized third and fourth cumulants of the distribution of the data; $\gamma_1 = \gamma_2 = 0$ for normally distributed data, while for symmetrically distributed data $\gamma_1 = 0$ and γ_2 is the kurtosis. For normally distributed data the efficiency of maximum likelihood with respect to generalized least squares is $(1 + 2\sigma^2)$, which for $\sigma = 1.0$ means that the maximum-likelihood estimate is three times more efficient. In many problems with constant coefficient of variation and nearly normal distributions we have found $\sigma \leqslant 0.35$ with a typical value $\sigma \simeq 0.20$. In these cases, the maximum increase in efficiency is 25% and the typical increase in efficiency for normally distributed data is 8%. One danger of blindly and routinely using normal-theory maximum likelihood can be seen by what happens at distributions that are not normal. For example, if $\gamma_1 = 0, \gamma_2 = 2.2$, and $\sigma = 0.20$, then the normal-theory maximum-likelihood estimate is already less efficient than generalized least squares. If the data were distributed as a gamma random variable with constant coefficient of variation $\sigma = 0.20$, then $\gamma_1 = 0.40$, $\gamma_2 = 0.24$ and the normal-theory maximum-likelihood estimate is 7% less efficient than generalized least squares. If $\sigma = 0.35$, rather than the 25% increase in efficiency predicted at the normal model, at the gamma model the normal-theory maximum-likelihood estimate is 18% less efficient, which simply reiterates our main point.

Since in practice one cannot easily distinguish between normal and gamma distributions with coefficient of variation $\sigma = 0.20$, using normal-theory maximum likelihood even when the data appear to be nearly normally distributed can be inefficient.

Besides the fact that normal-theory maximum likelihood may not be any more efficient than generalized least squares, a second danger arises if the variance function is not specified correctly. Suppose that, instead of having constant coefficient of variation σ, the standard deviation of the responses is actually $\sigma\mu^{1+\theta}$. The generalized least-squares estimate of μ is still $\bar{y}$, and in particular it still consistently estimates μ. The maximum-likelihood estimate assuming constant coefficient of variation σ converges not to μ but rather to

$$\mu[(1 + 4\sigma^2 + 4\sigma^4\mu^{2\theta})^{1/2} - 1.0]/(2\sigma^2)$$

To summarize this simple example, if the mean and variance functions are correctly specified and the data are actually normally distributed, then maximum likelihood will yield improved estimates over generalized least squares. However, the size of the improvement may not be very large and is likely to be less than 25% in practice. Maximum likelihood is less robust than generalized least squares, in that it is more adversely affected by incorrect variance-function specification and nonnormality in the direction of either positive skewness or kurtosis. Even data that may appear normally distributed may have sufficient skewness and kurtosis to make the maximum-likelihood estimate inefficient when compared to generalized least squares.

All these points carry over to regression models. The possible gains in efficiency are often less than striking and will often be losses. If the parameter θ in our model (2.1) is incorrectly specified, then, while generalized least squares is consistent, normal-theory maximum likelihood will not even consistently estimate β. Finally, other statisticians have asserted to us their belief that even under ideal circumstances maximum likelihood attains its asymptotic distribution less rapidly than does generalized least squares. We tend to believe this last conjecture, but await further study of the issue.

2.5   Inference about the regression parameter

Most statistical packages allow the user to specify estimated weights and perform a weighted nonlinear least-squares fit to the data. Often, the weights are known, e.g., the inverse of the number of replicates at each point. At other times the weights are unknown and must be estimated. In this section, we review some of the common methods for making inference about the regression parameter β when the method of estimation is generalized least squares.

A common analysis rests on the basic Theorem 2.1. For large enough sample sizes, the generalized least-squares estimate $\hat{\beta}_G$ is normally distributed with mean β and covariance matrix $(\sigma^2/N)S_G^{-1}$ (see equation (2.4)). Referring back to section 2.2, let $\hat{\beta}_*$ be the preliminary estimator of β used to form the weights (2.6). If one is iterating, $\hat{\beta}_*$ is the final generalized least-squares estimate. Many statistical packages ask the user to input these weights, so that estimates of σ^2 and S_G can be formed from the data through the formulae

$$\hat{\sigma}_G^2 = (N - p)^{-1} \sum_{i=1}^{N} \{[y_i - f(x_i, \hat{\beta}_G)]/g(\mu_i(\hat{\beta}_*), z_i, \theta)\}^2 \quad (2.14)$$

$$\hat{S}_G = N^{-1} \sum_{i=1}^{N} f_\beta(x_i, \hat{\beta}_G) f_\beta(x_i, \hat{\beta}_G)^T / g^2(\mu_i(\hat{\beta}_*), z_i, \theta) \quad (2.15)$$

Asymptotic standard errors for $\hat{\beta}_G$ are given by the square root of the diagonal elements of $(\hat{\sigma}^2/N)\hat{S}_G^{-1}$, and from these quantities asymptotic confidence intervals can be formed and hypothesis tests made. For example, the usual symmetric confidence interval for a scalar parameter β_2 is

$$\hat{\beta}_2 \pm t_{(1-\alpha/2)}^{N-p} \text{ Standard error } (\hat{\beta}_2)$$

where $t_{(1-\alpha/2)}^{N-p}$ is the $(1 - \alpha/2)$ percentile of a t-distribution with $N - p$ degrees of freedom. The standard error of $\hat{\beta}_2$ is the square root of the appropriate diagonal element of $(\hat{\sigma}_G^2/N)\hat{S}_G^{-1}$, which can be read off from computer output.

Since the asymptotic standard errors given above are standard features of statistical packages, they tend to be used to make inference about the regression parameter β. This is called a Wald-type inference after Wald (1943) (see Chapter 7). Even when the variances are constant, Bates and Watts (1980, 1986), Bates et al. (1982), and Ratkowsky (1983) have emphasized that Wald-type inferences can be inaccurate because they rely so heavily upon the linear approximation

$$f(x, \hat{\beta}_G) \simeq f(x, \beta) + f_\beta(x, \beta)^T(\hat{\beta}_G - \beta) \quad (2.16)$$

where as usual f_β is vector of partial derivatives of f. These authors provide methods for measuring 'curvature'. Large curvature means that inference based on the linear approximation (2.16) is potentially troublesome. The assessment of parameter-effects curvature, the component of the curvature that can be removed by reparameterization, is discussed in detail by Bates and Watts (1988). If the linear approximation is really questionable, typical advice is either to use likelihood ratio tests (see below) or to reparameterize the model. Appropriately parameterizing a nonlinear regression model to avoid problems with parameter-effects curvature remains largely an art.

The major advantage of Wald inference is that it is easy to perform and is widely available. It is instructive to compare three alternatives, two based on likelihood and one based on a resampling scheme called the bootstrap. The first method is based on the likelihood ratio test, and is 'nonparametric' in the same way as generalized least squares. Consider the redefined model (2.2), with estimated weights (2.6). The redefined model is approximately homoscedastic in its variances, so that one can compute ordinary likelihood ratio tests. For example, write

$$\beta^T = (\beta_1^T, \beta_2^T)$$

where β_1 is a vector of length p and β_2 is a vector of length r. Suppose we wish to test the null hypothesis

$$H_0: \qquad \beta_2 = \beta_{2,0}$$

where $\beta_{2,0}$ is specified. Fix the estimated weights based on the full model and let $\hat{\beta}_{G,0}$ be the generalized least-squares estimate of β assuming that the null hypothesis is true and $\beta_2 = \beta_{2,0}$. If we define

$$L(\beta) = -(N/2)\log\left(\sum_{i=1}^{N} [y_i^* - f^*(\hat{w}_i, x_i, \beta)]^2\right)$$

$$= -(N/2)\log\left(\sum_{i=1}^{N} \hat{w}_i[y_i - f(x_i, \beta)]^2\right)$$

then the likelihood ratio test rejects the null hypothesis if

$$-2[L(\hat{\beta}_{G,0}) - L(\hat{\beta}_G)] \geqslant \chi_r^2(1 - \alpha)$$

where $\chi_r^2(1 - \alpha)$ is the $(1 - \alpha)$th percentile of a chi-squared distribution with r degrees of freedom. Letting SSE_G be the weighted sum of squares

$$\sum_{i=1}^{N} \hat{w}_i[y_i - f(x_i, \hat{\beta}_G)]^2$$

and defining $SSE_{G,0}$ to be the corresponding sum of squares evaluated at the null-hypothesis estimate $\hat{\beta}_{G,0}$, we see that the likelihood ratio test rejects the null hypothesis if

$$N[\log(SSE_{G,0}) - \log(SSE_G)] \geqslant \chi_r^2(1 - \alpha)$$

An asymptotically equivalent version of this test can be formed by analogy with the ordinary F-test in linear regression and is to reject the hypothesis when

$$F = \frac{(SSE_{G,0} - SSE_G)/r}{SSE_G/(N-p)} \geqslant F(r, N-p, 1-\alpha) \qquad (2.17)$$

where $F(r, N-p, 1-\alpha)$ is the $(1-\alpha)$ percentage point of the F-distribution with r and $N-p$ degrees of freedom. The major computational difference between the F-test and the Wald test is that the former requires only one weighted least-squares minimization, while the latter requires as many minimizations as there are parameters to be tested.

Computing confidence regions for components β_2 of the entire regression parameter β is straightforward in principle. The standard $100(1-\alpha)\%$ confidence region for β_2 is the set of all $\beta_{2,0}$ for which the inequality (2.17) is not satisfied, i.e., the set where the null hypothesis is accepted. Computing such a region requires repeated nonlinear minimizations. In contrast, the usual Wald interval for a scalar parameter β_2 discussed earlier is the easily computed but possibly less accurate

$$\hat{\beta}_2 \pm t_{(1-\alpha/2)}^{N-p} \text{ Standard error } (\hat{\beta}_2)$$

A second approach to testing which is of approximately the same computational difficulty as likelihood ratio tests is the method of quasi-likelihood (see Wedderburn, 1974; McCullagh, 1983; McCullagh and Nelder, 1983). A motivation for defining a quasi-likelihood is to extend the definition of generalized nonlinear models (2.11) to allow fitting with an arbitrary variance function. In practice, quasi-likelihood estimates of β are generalized least-squares estimates based on solving (2.7). The quasi-likelihood function (an analog to the loglikelihood function) is the solution l_Q to the equation

$$\frac{\partial l_Q(\mu_i, y_i, \theta)}{\partial \mu_i} = \frac{y_i - \mu_i}{g^2(\mu_i, z_i, \theta)} \qquad (2.18)$$

The quasi-likelihood for the entire data set is

$$L_Q(\beta) = \sum_{i=1}^{N} l_Q(\mu_i(\beta), y_i, \theta)$$

and the maximum of the quasi-likelihood is attained at the solution to equation (2.7). The standard suggestion is to treat quasi-likelihoods as if they were actual loglikelihoods. Thus, the quasi-likelihood test statistic for the previous hypothesis is to reject the null hypothesis when

$$-2[L_Q(\hat{\beta}_{G,0}) - L_Q(\hat{\beta}_G)]/\hat{\sigma}_G^2 \geq \chi_r^2(1-\alpha) \tag{2.19}$$

The existence of a solution to (2.18) is known for many important examples. The power-of-the-mean model (2.5) is a quasi-likelihood model with the choice

$$l_Q(\mu, y, \theta) = yh(\mu, 1 - 2\theta) - h(\mu, 2 - 2\theta) \tag{2.20}$$

where $h(\mu, \lambda)$ is the Box and Cox (1964) power transformation

$$h(\mu, \lambda) = \begin{cases} (\mu^\lambda - 1)/\lambda & \text{if } \lambda \neq 0 \\ \log(\mu) & \text{if } \lambda = 0 \end{cases}$$

With somewhat more work, in the case that the variances are linear or quadratic in the mean one can construct a quasi-likelihood as well.

A fourth method of inference is based on the bootstrap; see Freedman and Peters (1984), Efron and Tibshirani (1986), Wu (1986), Efron (1987), and Hall (1988) for discussions. While computationally intensive, bootstrapping has potential use for problems of moderate sample sizes especially. Write the standardized errors as

$$\varepsilon_i = [y_i - f(x_i, \beta)]/[\sigma g(\mu_i(\beta), z_i, \theta)] \tag{2.21}$$

One often sees the assumption being made that the errors are independent and identically distributed, but we believe that this is not so crucial in the particular instance. How crucial this assumption is remains a matter of debate; see the discussion of Wu's (1986) article. The idea of the bootstrap is to form estimates of the errors

$$r_i = r_i(\hat{\beta}_G, \theta) = [y_i - f(x_i, \hat{\beta}_G)]/g(\mu_i(\hat{\beta}_G), z_i, \theta)$$

and then resample from the finite residual error distribution. One specific suggestion is to sample with replacement from the set $(r_1, \ldots, r_N)$, obtaining the bootstrap errors $(\varepsilon_1^b, \ldots, \varepsilon_N^b)$. Alternatively, the bootstrap errors could be generated from a smoothed instead of the empirical residual error distribution; Professor O. Bunke has told us of recent theoretical research showing that such smoothing improves the accuracy of the bootstrap. In either case, one might consider the use of the studentized residuals. The bootstrap sample observations are generated from the model

$$y_i^b = f(x_i, \hat{\beta}_G) + g(\mu_i(\hat{\beta}_G), z_i, \theta)\varepsilon_i^b \qquad i = 1, \ldots, N$$

We then estimate the regression parameter by generalized least squares applied to the bootstrap data set, obtaining $\hat{\beta}_{G,b}$, $\hat{\sigma}_{G,b}$, and $\hat{S}_{G,b}$. This procedure is to be repeated a large number of times, so that we have a sequence of bootstrap estimates

$$\{\hat{\beta}_{G,b,k}, \hat{\sigma}_{G,b,k}, \hat{S}_{G,b,k}\} \qquad k = 1, 2, \ldots, M$$

The bootstrap covariance estimate for generalized least squares is the sample covariance of the bootstrap estimates, i.e.

$$M^{-1} \sum_{k=1}^{M} (\hat{\beta}_{G,b,k} - \hat{\beta}_G)(\hat{\beta}_{G,b,k} - \hat{\beta}_G)^{\mathrm{T}} \qquad (2.22)$$

which replaces the asymptotic estimate $(\hat{\sigma}_G^2/N)\hat{S}_G^{-1}$. Wu (1986) has suggested multiplying the bootstrap covariance estimate by $N/(N - p)$. In principle, one can perform Wald inference using the bootstrap covariance estimate (2.22). Alternatively, one could bootstrap the ordinary F- and t-statistics to develop percentiles based on the data rather than the asymptotic normal distribution of generalized least squares. See Abramowitz and Singh (1985), Efron and Tibshirani (1986), Beran (1986), Efron (1987), and especially Hall (1988) for further discussion.

There are many potential advantages to using the bootstrap, one of which is the following. We have seen in section 2.3 that the covariance matrix of $N^{1/2}(\hat{\beta}_G - \beta)$ can be written as

$$\sigma^2 S_G^{-1} + N^{-1} V_N$$

(see equation (2.8)). Carroll *et al.* (1987) show that while the usual estimate is $\hat{\sigma}^2 \hat{S}_G^{-1}$, the bootstrap estimate is

$$\hat{\sigma}_G^2 \hat{S}_G^{-1} + N^{-1} V_N + (\text{smaller-order terms})$$

Thus, the bootstrap covariance estimate takes the usual estimate and automatically corrects for the second term in the variance expansion (2.8).

2.6 General covariance structures

Only notational changes are necessary to cope with the case of a general covariance matrix $\Lambda(\mu(\beta), Z, \theta)$ with θ known. For the generalized least-squares algorithm of section 2.2, combine steps 2 to 4 so that $\hat{\beta}_G$ minimizes in β

$$[Y - \mu(\beta)]^{\mathrm{T}} \hat{\Lambda}^{-1} [Y - \mu(\beta)]$$

where
$$\hat{\Lambda} = \Lambda(\mu(\hat{\beta}_*), Z, \theta)$$

and $\hat{\Lambda}^{-1}$ is a generalized inverse of $\hat{\Lambda}$. Fully iterating generalized least squares is equivalent to finding the solution to the following version of (2.7)

$$0 = D(\mu(\beta))^{T}[\Lambda^{-1}(\mu(\beta), Z, \theta)][Y - \mu(\beta)]$$

where $D(\mu(\beta))$ is the $(N \times p)$ matrix of partial derivatives of $\mu(\beta)$.

Theorem 2.1 also continues to hold with the asymptotic covariance matrix of $\hat{\beta}_G$ being $(\sigma^2/N)S_G^{-1}$, where

$$S_G(\beta) = N^{-1}D(\mu(\beta))^{T}[\Lambda^{-1}(\mu(\beta), Z, \theta)]D(\mu(\beta))$$

Wald-type inference is based on the estimated standard errors formed from the diagonals of the matrix

$$[\hat{\sigma}_G^2(\hat{\beta}_G)/N]S_G^{-1}(\hat{\beta}_G)$$

where
$$\hat{\sigma}_G^2(\beta) = (N - p)^{-1}[Y - \mu(\beta)]^{T}\hat{\Lambda}^{-1}[Y - \mu(\beta)]$$

Normal-theory likelihood ratio tests are as in (2.17). For quasi-likelihood inference, equation (2.18) now becomes

$$\frac{\partial l_Q(\mu, Y, \theta)}{\partial \mu} = \Lambda^{-1}(\mu(\beta), Z, \theta)[Y - \mu(\beta)]$$

2.7 Plotting techniques

There are a number of graphical techniques that can be used to detect heterogeneity of variance and to help determine a model for the heterogeneity. Throughout this section, we will assume that the variances follow the basic mean and variance model (2.1) with θ known.

The most widely used diagnostic for heterogeneity is the unweighted least-squares residual plot. Predicted values $\hat{y}_i = f(x_i, \hat{\beta})$ from an unweighted fit are plotted along the horizontal axis, while the residuals from the fit $y_i - \hat{y}_i$ are plotted along the vertical axis. A fan-shaped pattern in the plot indicates that residual variability depends on the mean response. Often there is value in plotting the residuals against predictor variables and the logarithm of the predicted values. Deviance and Anscombe residuals (Anscombe, 1961) for generalized linear models could also be used, although they require a specific distributional form (see McCullagh and Nelder, 1983; Pierce and Schafer, 1986).

The inadequacies of the standard plot dictate supplementation. It is not the best plot for detecting heterogeneity of variance (see below). Further, it gives no information to help model the variability.

Cook and Weisberg (1982) argue for plotting squared residuals against the predicted values, stating that 'the usual plots... are often sparse and difficult to interpret, particularly when the positive and negative residuals do not appear to exhibit the same general pattern. This difficulty is at least partially removed by plotting (squared residuals)... and thus visually doubling the sample size'. A wedge-shaped pattern would indicate heteroscedasticity increasing with the mean. In those problems with a moderately large residual, squaring can cause scaling problems for automatic plotting routines. Since this phenomenon is more the rule than the exception, we have found it useful instead to plot transformations of absolute residuals. This might be the absolute residuals themselves, their logarithms, or their 2/3 power. If the data were actually normally distributed with constant variance, the cube root of the squared residuals is the Wilson–Hilferty transformation to normalize variables with a chi-squared distribution. While squared residuals can be thought of as estimates of the variance, absolute residuals can be thought of as estimating relative standard deviation.

Our major point here is that understanding the structure of the variability can be facilitated by analysis of absolute residuals and transformations of them. For this reason we think of the absolute residuals as the basic building blocks in the analysis of heteroscedasticity.

We have often found it useful to plot the logarithm of the absolute residuals against the logarithm of the predicted values, the former being a proxy for the logarithm of the standard deviation and the latter a proxy for the logarithm of the mean. If the power-of-the-mean model (2.5) holds, then this plot occasionally looks linear with a slope related to θ.

A disadvantage of squared residuals is that their distribution is often markedly skewed, and hence they may fail to have the properties possessed by normal residuals. For example, if we assume that the errors are approximately normally distributed, then squared residuals will have an approximate chi-squared distribution with one degree of freedom. In this case, as discussed by McCullagh and Nelder

(1983, pp. 29 and 30), the cube root of the squared residual, being the cube root of a chi-square random variable, will have a distribution much closer to that of a normal random variable. Cook and Weisberg (1982) suggest further refinements to the basic residual plot. While their work applies formally only to linear regression, it is easily extended to nonlinear models. We first give the linear-model argument. Write the linear model as

$$E(Y) = X\beta$$

where X is the $(N \times p)$ design matrix. The predicted value vector is $\hat{Y} = HY$, where

$$H = X(X^T X)^{-1} X^T$$

is the $(N \times N)$ hat matrix with elements (h_{ij}). The individual residuals have mean zero and variances

$$\text{Variance}(r_i) = (1 - h_{ii})^2 \sigma_i^2 + \sum_{k \neq i} \sigma_k^2 h_{ik}^2$$

where σ_i is the standard deviation of y_i. Even if the data were homoscedastic with constant variance σ^2, since H is idempotent the residuals have variance proportional to $(1 - h_{ii})$. If the values of the leverages h_{ii} are closely related to the means and predicted values, the ordinary residual plots would then show a pattern of heteroscedasticity caused entirely by the design. Thus it makes good sense to replace in the basic plots the ordinary residuals by the studentized residuals, defined as

$$\text{Studentized residuals} = b_i = r_i / [\hat{\sigma}(1 - h_{ii})^{1/2}] \qquad (2.23)$$

Studentized residuals are available on almost every statistical package for linear regression.

The reasoning behind Cook and Weisberg's second modification is more involved. Its major purpose is to improve our ability to detect heteroscedasticity by slightly changing what is plotted along the horizontal axis. Suppose that the variances satisfy model (2.1) with θ a scalar, where by convention $\theta = 0$ implies constant variance σ^2. Defining

$$s_i = \frac{\partial}{\partial \theta} [g^2(\mu_i(\beta), z_i, \theta)]|_{\theta=0} \qquad (2.24)$$

we have that for small θ

$$\text{Variance}(\hat{\sigma}b_i) = (1 - h_{ii})\sigma_i^2 + \sum_{i \neq k} \sigma_k^2 h_{ik}^2/(1 - h_{ii})$$

$$\simeq \sigma^2 \left[1 + \theta\left((1 - h_{ii})s_i + \sum_{i \neq k} s_k h_{ik}^2/(1 - h_{ii}) \right) \right]$$

When all the off-diagonal elements h_{ik} are small, we can make the further approximation

$$\text{Variance}(b_i) \simeq 1 + \theta(1 - h_{ii})s_i$$

Their suggestion is then to plot ordinary, absolute, and squared studentized residuals b_i against $(1 - h_{ii})s_i$.

It is interesting to note the form of s_i in important cases. In the exponential-variance model

$$\text{Variance}(y_i) = \sigma^2 \exp\left[2\theta f(x_i, \beta)\right]$$

we have $s_i = f(x_i, \beta)$, while for the power-of-the-mean model (2.5) we obtain $s_i = \log[f(x_i, \beta)]$. Thus, plotting studentized residuals against predicted values will be most appropriate for detecting an exponential-variance model, while plotting against the logarithms of the predicted values might be used to detect whether the variances are a power of the mean. The same qualitative idea holds if we plot the residuals against any single predictor and any function of the predictors.

In nonlinear regression, an explicit formula for the variances of least-squares residuals is not available, although there is a version of the 'hat' matrix which can be used. In linear regression, the residual vector can be written as

$$\text{Residual vector} = (I - H)(Y - \mu)$$

Let $\mu(\beta)$ be the vector with ith element $f(x_i, \beta)$, and let X_* be the $(N \times p)$ matrix with ith row $f_\beta(x_i, \hat{\beta}_L)^T$. Then for unweighted least squares we have the approximations

$$\hat{\beta}_L - \beta \simeq (X_*^T X_*)^{-1} X_*^T [Y - \mu(\beta)] \tag{2.25}$$

$$\mu(\hat{\beta}_L) \simeq \mu(\beta) + X_*(\hat{\beta}_L - \beta) \tag{2.26}$$

If we define the nonlinear hat matrix

$$H = H(\beta) = X_*(X_*^T X_*)^{-1} X_*^T \tag{2.27}$$

with elements (h_{ij}), then the nonlinear least-squares residuals satisfy the approximation

$$Y - \mu(\hat{\beta}_L) = Y - \mu(\beta) - [\mu(\hat{\beta}_L) - \mu(\beta)] \simeq [I - H(\beta)][Y - \mu(\beta)]$$

$$(2.28)$$

Thus, the nonlinear hat matrix takes the place of the linear regression hat matrix. If h_{ii} is the ith diagonal element of the estimated nonlinear hat matrix $H(\hat{\beta}_L)$, the studentized residuals become

$$b_i = r_i/[\hat{\sigma}(1 - h_{ii})^{1/2}]$$

$$(2.29)$$

Further discussion and improved approximations are given by Cook and Tsai (1985).

For the heteroscedastic regression model (2.1), we assume that full iteration of the generalized least-squares algorithm has been used to compute the estimate $\hat{\beta}_G$. Let the estimated weights be

$$\hat{w}_i = g^{-2}(\mu_i(\hat{\beta}_G), z_i, \theta)$$

Then, as noted in (2.2) and (2.3), $\hat{\beta}_G$ can be computed as the ordinary unweighted nonlinear least-squares estimate on the redefined model

$$y_i^* = y_i \hat{w}_i^{1/2}$$
$$f^*(\hat{w}_i, x_i, \beta) = f(x_i, \beta)\hat{w}_i^{1/2}$$

The usual procedure (McCullagh and Nelder, 1983) is to ignore the estimation of the weights and to use the studentized residuals from the least-squares fit to the redefined model. These 'studentized residuals' have the form (2.29), with

$$r_i = [y_i - f(x_i, \hat{\beta}_G)]/g(\mu_i(\hat{\beta}_G), z_i, \theta)$$

$$(2.30)$$

where

$$X_* = (N \times p) \text{ matrix with } i\text{th row } f_\beta(x_i, \hat{\beta}_G)^{\mathrm{T}}/g(\mu_i(\hat{\beta}_G), z_i, \theta)$$

and

$$H(\hat{\beta}_G) = X_*(X_*^{\mathrm{T}}X_*)^{-1}X_*^{\mathrm{T}}$$

It is interesting to study the common studentization in a little more detail. Let σE be the vector with ith element

$$[y_i - f(x_i, \hat{\beta}_G)]/g(\mu_i(\hat{\beta}_G), z_i, \theta)$$

Let V_β be the $(N \times p)$ matrix with ith row

$$(\partial/\partial\beta^{\mathrm{T}})\log[g(\mu_i(\beta), z_i, \theta)]$$

Then

$$\hat{\beta}_G - \beta \simeq (X_*^T X_*)^{-1} X_*^T \sigma E$$

and the vector R of residuals with ith element (2.30) can be expanded as

$$R \simeq \sigma(I - H)E - \sigma^2 E \times [V_\beta(X_*^T X_*)^{-1} X_*^T E] \qquad (2.31)$$

where here '$\times$' means element-by-element multiplication. The usual studentization ignores the second term in (2.31). This will be acceptable in many circumstances, especially when σ is small.

There are difficulties that need to be kept in mind when interpreting residual plots, the first being the effect of varying degrees of data density along the horizontal axis. For example, suppose that one segment on the horizontal axis has 100 data points and a second segment of equal length has only five such points. The absolute residuals in the first segment will tend to look more dispersed even if there is no heterogeneity of variance, simply because the first segment will include a few large values which will draw visual attention. If, as is often the case, the data density is related to the mean value, then the effect can be to have the absolute residual plots indicate heterogeneity when no such heterogeneity exists.

When confronted with a residual plot based on a fairly large number of observations and varying degrees of data density, there are a number of ways of quickly assessing heterogeneity of variance. Often, the varying data density is alleviated by plotting the logarithm of the predicted values along the x-axis, since many designs are logarithmic in the mean. For a quick numerical indication, we have found it useful to compute the Spearman rank correlation coefficients of the absolute studentized residuals with the variables plotted along the horizontal axis. This correlation is computed by first ordering each variable from largest to smallest and replacing the numerical outcome by its rank, i.e., the smallest gets rank 1, the next rank 2, etc. The Spearman rank correlation is the ordinary correlation between the two sets of ranks. The Spearman correlation is not changed by monotone transformations, so that squared residuals or their logarithms could be used, as well as the logarithms of predictors or the predicted values. Positive coefficients would indicate increasing variances. The qualitative impression one takes from a residual plot should correspond to the sign of the rank correlation. The significance levels that accompany the Spearman correlation may be taken as rough indicators. Since the residuals and predicted values depend on estimated parameters, there is no formal theory and these significance levels should be interpreted with some care.

A second way to cope with the density effect in large data sets while at the same time obtaining information about a model for the variability is to use nonparametric regression techniques. See Silverman (1985) for a recent review and Carroll (1982a) and Cohen et al. (1984) for examples applied to heteroscedastic linear models. The idea here is to estimate the expected value of transformations of the absolute residuals as a function of the mean, in much the same way that one uses the ordinary scatterplot to help understand the structure of the means. The absolute, log-absolute, and cube root of the squared residual plots form the basic pictures for understanding the variance structure and can be thought of as analogs to a scatterplot. We have found it useful to fit a smooth function to various transformations of the absolute residuals, the function giving us an idea of the behavior of the standard deviations. When using logarithms of absolute residuals, one needs to be careful not to overinterpret the almost inevitable few very negative values caused by a small number of very small absolute residuals. Typically, we delete the smallest few percent of the absolute residuals before taking logarithms. In the terminology of Chapter 6, this is a crude way of bounding one part of the influence function.

One device for examining the structure of the standard deviations is to group the data into sets of equal size, the ordering being done on the horizontal axis, and then to plot the mean transformed absolute residual in each group against the corresponding mean horizontal axis value for that group. This device can be surprisingly effective. Slightly more sophisticated smoothing can be done by taking moving averages, medians or trimmed means. Depending on one's software, one can become as sophisticated as using spline or kernel smoothing as discussed in Silverman (1985). The smoother we used in the following examples was a kernel regression estimator based on the uniform kernel, the bandwidth being chosen by trial and error. Endpoint effects were adjusted for by selective deletion. For more formal details of kernel regression, see also Chapter 3.

A second problem with unsmoothed residual plots occurs when a small percentage of the data are poorly fit for a few large or small predicted values, particularly in a density-dependent plot. A casual glance at the plot will often suggest severe heterogeneity which is difficult to justify upon closer examination. We typically cover the largest few absolute residuals with a piece of card; if the pattern remains suggestive then we can be more confident of our diagnosis. One might use robust smoothing as in Haerdle (1984), Haerdle and Gasser (1984), Haerdle and Tsybakov (1987) or Cohen *et al.* (1984). Other references on smoothing include Hall and Marron (1987), Haerdle *et al.* (1987), and Marron (1986, 1987).

Before looking at an example, it is worth noting that many residual plots show some heterogeneity of variance, and the question that will often arise is: 'Is there enough heterogeneity of variance so that I should go to the trouble of weighting?' To answer this question precisely requires the time and effort that the question implicitly seeks to avoid. As stated in the introduction, our attitude is that heteroscedasticity is the systematic and usually smooth change in variability as one perturbs the predictors, so that we will require not simply some heterogeneity but rather a discernible systematic pattern. A rough rule of thumb is that, if after smoothing the main body of the standard deviations increase by a factor of 1.5 or less, then probably weighting will not be crucial, but if this factor is 3.0 or more then weighting should be considered.

It is useful to study a specific example. The data set we use is the car insurance example (McCullagh and Nelder, 1983, p. 158). The data originally appeared in Baxter *et al.* (1980), with the original analysis done in an unpublished report by Baxter and Coutts in 1977. The data are average insurance claims as a function of three predictors: policy-holder's age group (PA), car group (CG), and vehicle age group (VA). This is a highly unbalanced three-way factorial with some missing cells and a random number of observations in each cell. The average insurance claims are based on between one and 434 claims, with the average number of claims in each cell being more than 70. While the original claim amounts are probably quite skewed, because of central-limit-theorem effects, the data used in the analysis will tend to be normally distributed. In general, there is some controversy as to whether one should analyze very skew data on the original scale by estimating means or by estimating medians, perhaps after a data transformation.

We use these data as a numerical illustration because they exhibit certain interesting features, but we do not claim a complete analysis. Indeed, any real data analysis is highly context-specific and claiming an improved analysis of someone else's data is usually as pointless as it is misleading. For example, it is clear that the number of claims in each cell is random and might reasonably be thought of as the most important component of the problem; Baxter *et al.* (1980) model claim frequency, although we do not. Even conditionally on the observed number of claims in each cell, it appears that in this example the variance structure is of far less interest than are the cell means. Because of the large number of observations, the mean model used in this analysis will have a considerably greater impact on the estimated cell means than will the variance structure.

Write y_{ijk} to be the observed average claim when $(PA, CG, VA) = (i, j, k)$, and let n_{ijk} represent the number of observations upon which y_{ijk} is based. We followed McCullagh and Nelder (1983) and fit a model in which the means were given by

$$E(y_{ijk}) = (\mu + \alpha_i + \beta_j + \gamma_k)^{-1}$$

Here α_i, β_j and γ_k refer to the usual analysis-of-variance effects. We took the standard deviation to follow the power-of-the-mean model (2.5) divided by the square root of the number of observations making up that observation.

In the analysis, all the published responses were divided by 1000. Figures 2.1–2.3 are plots of the cube root of the squared studentized residuals $|b_i|^{2/3}$ against the logarithm of the predicted values. The model (2.5) for the standard deviations was fit to these data, with $\theta = 0, 1.0$, and 1.5 in Figures 2.1, 2.2, and 2.3, respectively. If $\theta = 1.00$ were 'correct' we would expect Figure 2.1 to indicate constant variance and Figure 2.2 to have overcompensated, leading to a clear pattern of decreasing variance; opposite conclusions would emerge if $\theta = 1.50$.

It is important to remember that we have assumed a model for the means and are only trying to understand the structure of the variances. As with any model, one needs to be cautious about extrapolating the variance function, especially extrapolating the power-of-the-mean model to mean zero. The reader also needs to keep in mind that we are not comparing distributions, especially between the gamma distribution with $\theta = 1.00$ and the inverse Gaussian distribution, which is a generalized nonlinear model with $\theta = 1.50$.

The heterogeneity in Figure 2.1 is evident, the variability clearly increasing with the mean. This is very severe heteroscedasticity. The other two plots are somewhat more difficult to interpret because of

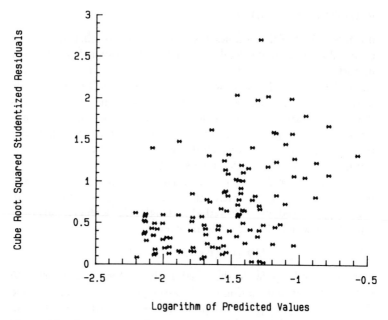

Figure 2.1   *Car insurance data: unweighted least-squares fit.*

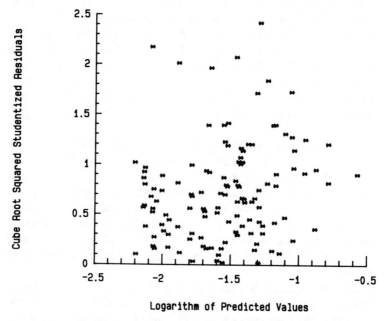

Figure 2.2   *Car insurance data: constant-coefficient-of-variation model.*

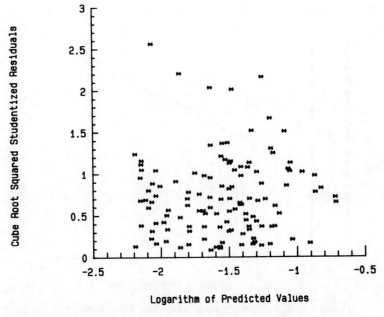

Figure 2.3   *Car insurance data: variance proportional to cube of mean.*

the large sample size, the four moderately large absolute residuals in
the center of each picture, and the variation in data density. The
smaller number of points with larger predicted values can lead one to
overinterpret a pattern of decreasing variances.

We asked 15 statisticians to study Figures 2.2 and 2.3 with a view to
understanding 'if there is sufficient heteroscedasticity of variance to
go to the bother of weighting, transforming, etc.'. Of these statisticians, only one stated that either picture was worrisome, and he
picked only Figure 2.2. We then provided these statisticians with the
Spearman rank correlation, this being $\rho = 0.22$ for Figure 2.2 with
formal significance level 0.01, while for Figure 2.3 the correlation is
$\rho = 0.06$ with significance level 0.54. We also sorted the data along
the horizontal axis into groups of approximately equal size and then
plotted the means of each group. With this information, no statistician singled out Figure 2.3 with $\theta = 1.50$ as potentially of concern,
but five suggested that Figure 2.2 with $\theta = 1.00$ would be a candidate
for weighting, the variances increasing with the mean. What is
interesting about this exercise is that even a group of experienced
statisticians do not all agree in their interpretation of these data.

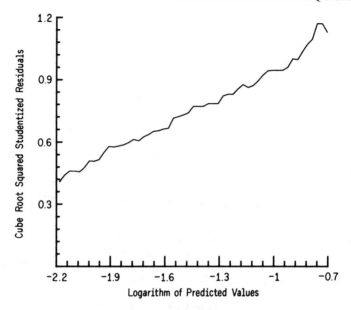

Figure 2.4   *Car insurance data: unweighted least-squares fit.*

In Figure 2.4 we display the smoothed cube root of the squared residuals with $\theta = 0$. A uniform kernel was selected with bandwidth 0.5, the horizontal axis selected to be from -2.2 to -0.8. Since these residuals are of the order of 0.4 on the left while of the order of 1.2 on the right, we see that the standard deviations seem to be increasing roughly by a factor of 3:1 in the data. Notice how much more informative this smooth plot is than is the ordinary residual plot.

In Figure 2.5 we plot the smoothed cube root of the squared residuals for $\theta = 0$, 1.0, and 1.5. Some of our colleagues have argued that even further smoothing is necessary, although we feel that this picture gives the essential features of the data. One might reasonably infer that setting $\theta = 1.50$ more adequately controls for heterogeneity than does $\theta = 1.00$. Clearly, the difference between these two choices is less crucial than the difference with no weighting at all.

In Figure 2.6 we plot the logarithm of the absolute residuals against the logarithm of the predicted values for the unweighted least-squares fit. There is a pronounced linear trend. Since in problems of very large sample size this plot would have slope θ, it makes sense to compute a least-squares straight-line fit with the slope being an estimate of θ. This simple technique is discussed in the next chapter, with the

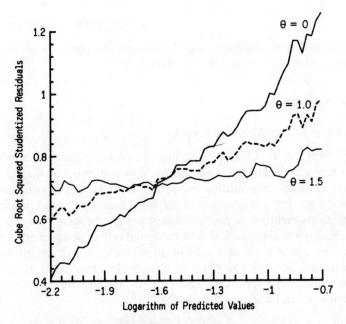

Figure 2.5   *Car insurance data.*

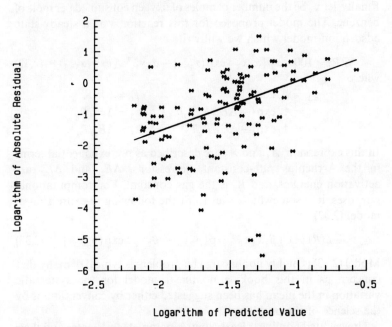

Figure 2.6   *Car insurance data: unweighted least-squares fit.*

variation that the five points with the smallest absolute residuals have
been deleted. For these data, the slope is 1.57.

2.8 Examples

Example 2.1   Oxidation of benzene

Pritchard et al. (1977) describe an interesting example concerning the
initial rate of oxidation of benzene over a vanadium oxide catalyst at
three reaction temperatures and several benzene and oxygen concentrations. For the purposes of graphical clarity we have removed
their thirty-eighth data point, although doing so does not affect our
qualitative conclusions. Let R be the initial rate of disappearance of
benzene; our R is the rate used by Pritchard et al. with their listed
numbers divided by 100. Let x_1 and x_2 be the oxygen and benzene
concentrations, respectively. Let T be the absolute temperature in
kelvins (K) and consider the variable

$$x_3 = 2000(1/T - 1/648)$$

Finally, let x_4 be the number of moles of oxygen consumed per mole of
benzene. The model proposed for this reaction was a steady-state
adsorption model which we will write as

$$E(R) = 100\alpha_1\alpha_2/[\alpha_1 x_2^{-1} \exp(\alpha_4\xi_3) + \alpha_2 x_1^{-1} x_4 \exp(\alpha_3\xi_3)] \quad (2.32)$$

where
$$\xi_3 = x_3/2000 \qquad\qquad T_0 = 648 \text{ K}$$
$$\alpha_1 = A_1 \exp(-\Delta E_1/R_g T_0) \qquad \alpha_3 = \Delta E_1/R_g$$
$$\alpha_2 = A_2 \exp(-\Delta E_2/R_g T_0) \qquad \alpha_4 = \Delta E_2/R_g$$

In this expression, A_1 and A_2 are described as pre-exponential terms
in the Arrhenius rate-constant expressions, ΔE_1 and ΔE_2 are
activation energies, and R_g is the gas constant. For computational
purposes, it is somewhat easier to fit the following re-expression of
model (2.32)

$$E(R) = 1/[\beta_1 x_4 x_1^{-1} \exp(\beta_3 x_3) + \beta_2 x_2^{-1} \exp(\beta_4 x_3)] \quad (2.33)$$

Model (2.33) is highly nonlinear. The problem is typical of many that
concern us in this book; a nonlinear model for the systematic
variation in the mean has been suggested either by convention or by
the science of the problem.

Unweighted nonlinear least-squares parameter estimates and their estimated standard errors (SE) are

$$\beta_1 = 0.97 \quad SE = 0.10 \qquad \beta_2 = 3.20 \quad SE = 0.20$$
$$\beta_3 = 7.33 \quad SE = 1.14 \qquad \beta_4 = 5.01 \quad SE = 0.64$$

The Spearman correlation between absolute studentized residuals from this least-squares fit and the predicted values is 0.53, with a significance level of less than 0.001. Such values indicate strong heterogeneity of variance in the model, and this is confirmed by the residual plots. In Figure 2.7 we plot the absolute studentized residuals against the logarithms of the predicted values; the wedge-shaped pattern is apparent. In Figure 2.8 we plot the logarithm of the absolute residuals against the logarithm of the predicted values, having deleted the two smallest of the former. There is a linear trend in this figure, and a least-squares fit gives a slope of 0.70, which as later analyses will suggest is probably an underestimate due to the initial unweighted fit to the means. It seems clear here that the variances increase with the mean response. The means vary systematically according to (2.32) and the variability varies systematically as a function of the mean response.

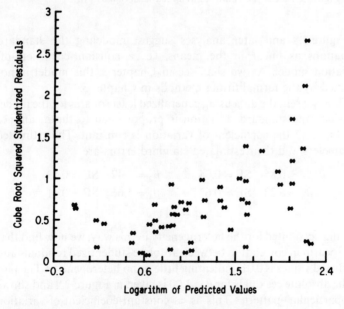

Figure 2.7 *Benzene data: unweighted least-squares residuals.*

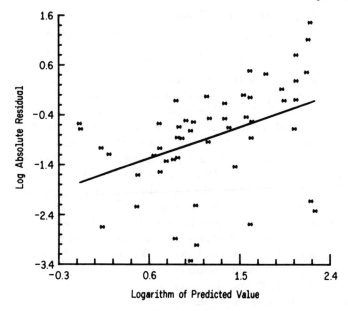

Figure 2.8   *Benzene data: unweighted least-squares fit.*

Figure 2.8 and later analyses suggest modeling the standard deviations as linear in the means, i.e., a constant-coefficient-of-variation model. As we shall see in Chapter 3, this model is not rejected by the formal fitting methods in Chapter 3.

Thus, we refit the data using generalized least squares for the model in which the standard deviation is proportional to the mean, i.e., $\theta = 1.00$ and the coefficient of variation is constant. The estimated parameters and their estimated standard errors are

$$\beta_1 = 0.86 \quad SE = 0.04 \qquad \beta_2 = 3.42 \quad SE = 0.13$$
$$\beta_3 = 6.20 \quad SE = 0.63 \qquad \beta_4 = 5.66 \quad SE = 0.41$$
$$\sigma = 0.11$$

Having accounted for the heterogeneity in this way, we now find that the Spearman correlation between absolute studentized residuals and predicted values is 0.01, indicating little or no heterogeneity. The plot of the absolute residuals for $\theta = 1.00$ is given in Figure 2.9 and shows no particular pattern. This is a constant-coefficient-of-variation model with estimated coefficient of variation 0.11.

The residual plots confirm that choosing $\theta = 1.00$ is much more

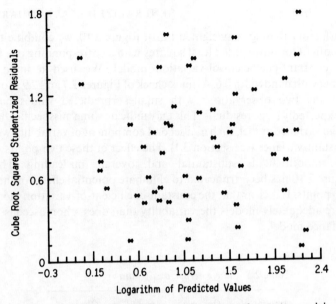

Figure 2.9 *Benzene data: constant-coefficient-of-variation model.*

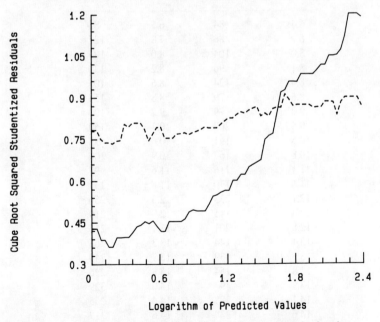

Figure 2.10 *Benzene data: unweighted, full line; weighted, broken line.*

satisfactory than an unweighted fit. In Figure 2.10, we combine the
smooth fit for unweighted least squares with a corresponding one for
the constant-coefficient-of-variation model. We used a uniform
kernel with bandwidth 0.6. A quick check of Figures 2.7 and 2.9 shows
that the two observations with smallest predicted values have
unexpectedly large residuals. This may indicate some misspecification
of the variance model, such as a second component of variability with
constant variance (see section 3.1). The effect of these two points on
the smooth fit is substantial and, overall, misleading. Thus
Figure 2.10 has been truncated to eliminate potential effects of these
two points. It is clear that the constant-coefficient-of-variation model
more adequately models the variability than does a homoscedastic-
variance model.

Table 2.3   *The esterase assay data*

Amount of esterase	Observed count	Amount of esterase	Observed count
6.4	84	6.5	85
6.7	86	7.8	127
8.0	104	8.0	107
8.1	96	8.2	130
8.4	124	8.6	105
8.6	79	8.8	153
9.0	79	9.2	100
9.5	203	10.3	159
10.5	191	10.6	93
10.8	167	10.9	100
11.1	116	11.6	97
11.6	170	11.8	131
12.1	233	12.3	256
12.6	115	12.8	219
12.8	201	13.1	215
13.1	144	13.3	268
13.3	139	13.7	249
13.8	154	13.8	226
13.9	97	14.0	329
14.1	288	14.4	255
14.6	239	14.6	317

(*Contd.*)

Table 2.3   (Contd.)

Amount of esterase	Observed count	Amount of esterase	Observed count
14.6	216	14.8	266
15.0	301	15.2	193
15.2	389	15.2	278
15.8	271	15.9	250
16.0	148	16.0	103
16.1	315	16.4	256
16.9	126	17.0	137
17.1	340	17.5	136
17.7	276	18.1	342
18.8	262	19.0	486
19.2	336	20.5	354
20.5	393	20.8	459
20.8	270	20.8	260
20.9	343	21.2	474
21.4	270	21.8	416
21.8	296	22.1	317
23.0	409	23.2	376
23.5	381	23.7	466
23.8	529	24.2	412
24.4	566	24.6	369
25.2	418	25.2	531
25.5	435	26.9	472
27.2	208	27.4	646
27.7	412	29.0	595
29.0	474	30.3	527
30.8	438	33.6	635
35.2	416	38.2	695
38.6	717	39.1	1042
40.5	695	40.9	239
41.2	597	41.7	1006
44.5	718	45.0	778
46.6	599	52.0	789
52.1	921	52.4	697
6.1*	52*	5.6*	166*
3.1*	28*	11.0*	423*
11.2*	373*		

*Those observations marked by an asterisk are deleted in the analysis of this chapter.

Example 2.2   Esterase count data

In Table 2.3, we present the results of an assay for the concentration of an enzyme esterase. The observed concentration of esterase was recorded, and then in a binding experiment the number of bindings were counted. These data are proprietary and we are unable to give further background. We do not know whether the concentration of esterase has been accurately measured, but for illustration we will assume little if any measurement error in this predictor. The lack of replicates in the response is rather unusual in our experience. Since the response is a count, one might expect Poisson variation, i.e., the power-of-the-mean model holds with $\theta = 0.50$. In our experience with assays, such a model almost always underestimates θ, with values between 0.60 and 0.90 being much more common (see Finney, 1976; Raab, 1981a). The complete data set is listed in Table 2.3. In the following analysis we deleted five points marked by an asterisk to make the illustration a little cleaner.

This is an example of a calibration experiment (see section 2.9). The eventual goal of the study is to take observed counts and infer the concentration of esterase, especially for smaller values of the latter. As

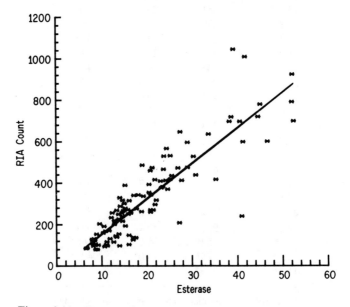

Figure 2.11   *Esterase data.*

is typical in these experiments, we measure a calibration data set where both the predictor variable, esterase, and the counted response are known. In Figure 2.11 we plot the observed data, which appear reasonably linear. Although the logarithm of the response plotted against the logarithm of the predictor also appears linear, and less heteroscedastic, we take as our model for the means in these data

$$f(x, \beta) = \beta_0 + \beta_1[\text{esterase}]$$

As is evident from Figure 2.11, the data exhibit severe heterogeneity of variance. The Spearman correlation between absolute studentized residuals and predicted values from an unweighted least-squares fit is $\rho = 0.39$ with formal significance level equal to or less than 0.0001. The estimated parameters are

$$\hat{\beta}_0 = -17.02 \qquad \hat{\beta}_1 = 17.05$$

The smoothed plot of absolute studentized residuals and predicted values as well as the logarithms of these quantities indicate that the constant-coefficient-of-variation model $\theta = 1.00$ is reasonable, although a value $\theta = 0.9$ might be slightly better. The log absolute

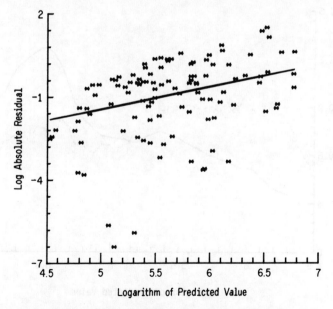

Figure 2.12 *Esterase data: unweighted least-squares fit.*

residual plot is given as Figure 2.12 and has least-squares slope 0.79, which is probably too small due to the unweighted fit to the mean. For $\theta = 1.00$, the Spearman correlation between absolute studentized residuals and predicted values is $\rho = -0.10$, with significance level 0.29. In Figure 2.13, we combine the smoothed plots for the unweighted fit $\theta = 0$ and the constant-coefficient-of-variation model $\theta = 1.0$; these used a uniform kernel with bandwidth 0.65. Some truncation of the plot for $\theta = 1.0$ was done. The residuals associated with the three smallest predicted values were quite small, which had an effect on the form of the final plot. As in the previous section, we chose to handle these end effects by truncation. Note that the choice $\theta = 1.0$ does a far better job of accounting for heteroscedasticity although it seems to have gone a little too far. For the constant-coefficient-of-variation model, the parameter estimates and their estimated standard errors are

$$\hat{\beta}_0 = -37.36 \quad SE = 12.21 \qquad \hat{\beta}_1 = 18.16 \quad SE = 0.95$$
$$\hat{\sigma} = 0.27$$

Once again, note the moderately small coefficient of variation.

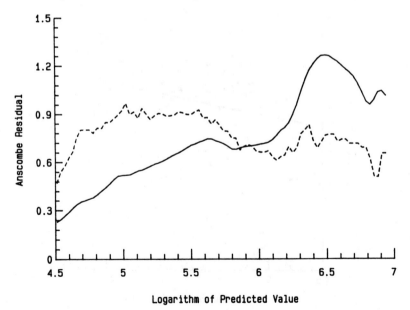

Figure 2.13 *Esterase data: unweighted, full line; weighted, broken line.*

In this example, we again see that simple plotting techniques help us understand the nature of the variability. However, we still do not have a particularly firm estimate of the power parameter θ, because our analysis has been based on unweighted least-squares residuals. The methods of estimating the variance function discussed in the next chapter are more satisfactory, since they use residuals from a weighted fit.

2.9 Prediction and calibration

One example where heterogeneity of variation occurs naturally is calibration experiments. In these experiments one first takes a training or calibration sample $(y_1, x_1), \ldots, (y_N, x_N)$ and fits models to the mean and variance structures. The real interest lies in an independent pair (y_0, x_0). In some instances x_0 is known and we wish to obtain point and interval predictors for y_0 and its expected value. In some calibration experiments, y_0 is easily measured but x_0 must be estimated. See Rosenblatt and Spiegelman (1981) for a general discussion, and Scheffe (1973), Lieberman et al. (1967), and Carroll et al. (1986c) for the case that there are more than one pair of unknowns. See also Garden et al. (1980) and Schwartz (1979) for applications.

In an assay x might represent the concentration of a substance and y might represent a counted value or intensity level that varies with concentration. One will have a new value y_0 of the count or intensity and wish to draw inference about the true concentration x_0. The calibration sample is drawn so that we have a good understanding of how the response varies as a function of concentration. The regression equation relating the response to concentration is then inverted to predict the concentration from the observed response. The area of calibration inference is currently under vigorous development, and our discussion can only be an introduction.

If we define the matrices

$$S_{L,1} = N^{-1} \sum_{i=1}^{N} f_\beta(x_i, \beta) f_\beta(x_i, \beta)^{\mathrm{T}}$$

and

$$S_{L,2} = N^{-1} \sum_{i=1}^{N} f_\beta(x_i, \beta) f_\beta(x_i, \beta)^{\mathrm{T}} g^2(\mu_i(\beta), z_i, \theta)$$

then the unweighted least-squares estimate $\hat{\beta}_L$ is asymptotically normally distributed with mean β and covariance matrix

$$(\sigma^2/N)S_{L,1}^{-1}S_{L,2}S_{L,1}^{-1} \qquad (2.34)$$

Since unweighted least squares consistently estimates β, in problems with a large sample size the actual numerical difference between $\hat{\beta}_L$ and the generalized least-squares estimate $\hat{\beta}_G$ may be rather small. To show that this idea is accepted outside the statistics literature, let us quote Schwartz (1979), who states 'there is one point of agreement among statistics texts and that is the minimal effect of weighting factors on fitted regression curves. Unless the variance nonuniformity is quite severe, the curve fitted to calibration data is likely to be nearly the same, whether or not the variance nonuniformity is included in the weighting factors'. This may lead one to question the importance of weighting, since the inefficiency of unweighted least squares in this instance is only relative. There are two compelling reasons for using generalized least squares even when the sample size of the calibration data set is large. The first has to do with inference about the regression parameters. While the actual covariance matrix of unweighted least squares is approximately (2.34), the estimated covariance matrix is approximately

$$(\sigma^2/N)N^{-1}\sum_{i=1}^{N} g^2(\mu_i(\beta), z_i, \theta)S_{L,1}^{-1} \qquad (2.35)$$

Thus the test statistics from an unweighted fit will not have the correct level, i.e., type I error; see for example Judge et al. (1985).

A second drawback to an unweighted fit is that it can lead to prediction or calibration confidence intervals which are either far too short or long. The size of the calibration data set, as long as it is moderately large, is usually relatively less important, with the form and estimation of the variance function assuming a primary role. For the remainder of this section we will take $z = x$ in the basic model (2.1) and assume that the errors

$$\varepsilon_i = [y_i - f(x_i, \beta)]/[\sigma g(\mu_i(\beta), x_i, \theta)] \qquad (2.36)$$

are normally distributed. It is possible to be somewhat more general and use bootstrap ideas to estimate the distribution function of the errors (2.36). This refinement will not be pursued here, being delayed until Chapters 4 and 5.

2.9.1 Prediction of a new response

Given a value x_0, the standard point estimate of the response y_0 is its mean $f(x_0, \beta)$. We use the term 'standard' here because the mean is not necessarily the best estimate for nonnormally distributed data. The variance in the error made by this prediction is

$$\text{Variance}(y_0 - f(x_0, \hat{\beta}_G)) \simeq \sigma^2 q_N^2(x_0, \beta, \theta) \qquad (2.37)$$

where

$$q_N^2(x_0, \beta, \theta) = g^2(f(x_0, \beta), x_0, \theta) + N^{-1} f_\beta(x_0, \beta)^T S_G^{-1} f_\beta(x_{0,\beta})$$

If the size N of the calibration data set is large, then the error in prediction is determined predominately by the variance function

$$\sigma^2 g^2(f(x_0, \beta), x_0, \theta)$$

and not by the calibration data set itself. An approximate $100(1 - \alpha)\%$ confidence interval for the response y_0 is given by

$$I(x_0) = \{\text{all } y \text{ in the interval } f(x_0, \hat{\beta}_G) \pm t_{1-\alpha/2}^{N-P} \hat{\sigma}_G q_N(x_0, \hat{\beta}_G, \theta)\}$$
$$(2.38)$$

where $t_{1-\alpha/2}^{N-P}$ is the $(1 - \alpha/2)$ percentage point of a t-distribution with $N - p$ degrees of freedom. For large sample sizes, this interval becomes

$$I(x_0) \simeq \{\text{all } y \text{ in the interval } f(x_0, \beta) \pm t_{1-\alpha/2}^{N-P} \sigma g(f(x_0, \beta), x_0, \theta)\}$$
$$(2.39)$$

The prediction interval (2.38) is only an approximate $100(1 - \alpha)\%$ confidence interval because the function q_N is not known and must be estimated.

The effect of ignoring the heterogeneity can be seen through examination of (2.39). If $\hat{\sigma}_L^2$ is the unweighted mean squared error, then for large samples we have the approximation

$$\hat{\sigma}_L^2 \simeq \sigma^2 g_{\text{mean}}^2 = \sigma^2 N^{-1} \sum_{i=1}^N g^2(\mu_i(\beta), x_i, \theta)$$

Thus the unweighted prediction interval for large sample sizes is approximately

$$I_L(x_0) \simeq \{\text{all } y \text{ in the interval } f(x_0, \beta) \pm t_{1-\alpha/2}^{N-P} \sigma g_{\text{mean}}\} \qquad (2.40)$$

Comparing (2.39) and (2.40) we see that, where the variability is small, the unweighted prediction interval will be too long and hence pessimistic, and conversely where the variance is large.

2.9.2 Confidence intervals for the mean response

If one wants a confidence interval for the mean $f(x_0, \beta)$ rather than the response itself, first write

$$q_{N*}^2 = N^{-1} f_\beta(x_0, \beta)^T S_G^{-1} f_\beta(x_0, \beta)$$

The approximate $100(1 - \alpha)\%$ confidence interval takes the form

$$f(x_0, \hat{\beta}_G) \pm \hat{\sigma} q_{N*} t_{1 - \alpha/2}^{N - p}$$

As with prediction, the validity of this interval depends on the choice of the variance function.

2.9.3 Calibration confidence intervals

Given the value of the response y_0, the usual estimate of x_0 is the set of all values x for which $f(x, \hat{\beta}_G) = y_0$. If the regression function is strictly increasing or decreasing, then the estimate of x_0 is that value which satisfies $f(x_0, \hat{\beta}_G) = y_0$. The most common confidence interval for x_0 is the set of all values x for which y_0 falls in the prediction interval $I(x)$, i.e.

Calibration interval for $x_0 = \{$all x with $y_0 \in I(x)$ in (2.38)$\}$

This interval has a certain intuitive charm, but in our experience it takes some effort to realize that it actually is a $100(1 - \alpha)\%$ confidence interval for the unknown concentration. The effect of not weighting is too long and pessimistic confidence intervals for x_0 where the variance is small and the opposite where the variance is large. As far as we know, little work has been done to determine whether one can shorten the calibration confidence interval by making more direct use of the variance function.

2.9.4 Straight-line regression

In the case of straight-line regression with an intercept, write

$$f(x, \beta) = \beta_0 + \beta_1 x \tag{2.41}$$

Suppose we have decided to use the weighting function $w(x)$ given by

$$w(x) = g^{-2}(f(x, \hat{\beta}_G), x, \theta)$$

It is important to note that we must specify a weighting function rather than just the weights if we are to do calibration. For prediction, the weighting function must be specified if prediction is to be done for values not in the original training sample. The need to specify a weighting function means in particular that merely obtaining estimated weights by replication in the training sample is not sufficient. To obtain explicit formulae for all quantities, we make the following definitions

$$a_i = w^{1/2}(x_i) \qquad b_i = x_i w^{1/2}(x_i)$$

$$c_i = y_i w^{1/2}(x_i) \qquad s_{aa} = \sum_{i=1}^{N} a_i^2$$

$$s_{bb} = \sum_{i=1}^{N} b_i^2 \qquad s_{ab} = \sum_{i=1}^{N} a_i b_i$$

$$\gamma = s_{aa} s_{bb} - s_{ab}^2$$

with s_{cc}, s_{ac}, and s_{bc} defined in the obvious manner. The generalized least-squares estimates are

$$\hat{\beta}_0 = (s_{bb} s_{ac} - s_{ab} s_{bc})/\gamma$$

$$\hat{\beta}_1 = (s_{aa} s_{bc} - s_{ab} s_{ac})/\gamma$$

$$\hat{\sigma}_G^2 = (N-2)^{-1} \sum_{i=1}^{N} w(x_i)[y_i - \hat{\beta}_0 - \hat{\beta}_1 x_i]^2.$$

Estimated standard errors for the regression parameters are

$$\text{Standard error } \hat{\beta}_0 = \hat{\sigma}_G (s_{bb}/\gamma)^{1/2}$$
$$\text{Standard error } \hat{\beta}_1 = \hat{\sigma}_G (s_{aa}/\gamma)^{1/2}$$

In this special case the prediction error variance is (2.37) with

$$q_N^2 = w^{-1}(x_0) + \gamma^{-1}(s_{bb} - 2s_{ab}x_0 + s_{aa}s_{new}^2)$$

The prediction interval (2.38) and its associated calibration interval use these values of $\hat{\sigma}_G$ and q_N, with the substitution (2.41) for $f(x, \beta)$.

In the constant-variance case, there is a closed-form expression for the calibration confidence interval based on solutions to quadratic equations (see Miller, 1981). For heterogeneous variability, no such closed-form solution exists in general. A simple way to obtain a calibration confidence interval from most statistical packages is to plot the regression line and the associated 95% prediction interval as a function of the predictor x, connecting the points by a smooth line. Draw a horizontal line through the value y_0, and project the intersection of this line with the upper and lower confidence band onto the x-axis. The interval so obtained is a calibration confidence interval for x_0.

2.9.5 Examples

For a numerical illustration, first consider Example 2.1 in section 2.8, the esterase assay data. As noted there, the constant-coefficient-of-variation model is reasonable for these data, with the estimated standard deviations varying by a factor of more than 6. The effect of not weighting should be to have prediction and calibration confidence intervals that are much too large for small amounts of esterase and conversely for large amounts. In Figure 2.14 we plot the 95%

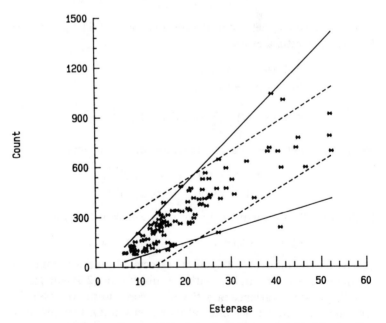

Figure 2.14 *Esterase assay: 95% prediction limits; unweighted, broken line; weighted, full line.*

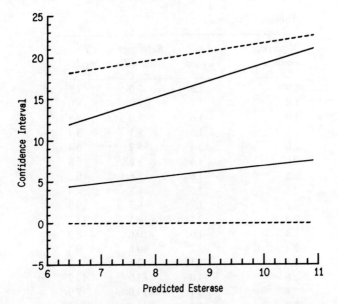

Figure 2.15   *Esterase assay: 95% calibration intervals; unweighted, broken line; weighted, full line.*

prediction intervals for the count response for unweighted versus weighted regression: the effect is clear. The size of the unweighted interval for the smaller values of esterase is almost ludicrous. In Figure 2.15 we plot the calibration confidence intervals for weighting and not weighting when the estimated esterase values range between 6.0 and 11.0; the difference between the weighted and unweighted intervals is substantial.

In Table 2.4 we list another data set for which calibration might be

Table 2.4   *Hormone assay data*

Reference method	Test method	Reference method	Test method
1.0	1.8	6.4	6.0
1.4	1.0	6.4	4.1
1.6	1.5	6.7	7.8
1.8	1.3	6.7	6.9
1.9	1.7	6.8	4.6
			(*Contd.*)

Table 2.4 (*Contd.*)

Reference method	Test method	Reference method	Test method
1.9	1.6	7.0	7.5
1.9	1.1	7.9	8.0
1.9	1.6	8.0	7.8
2.0	1.6	8.1	5.4
2.0	3.4	8.2	6.4
2.0	1.4	8.6	6.8
2.1	1.6	8.6	9.2
2.1	1.6	8.8	8.4
2.2	1.7	9.9	9.3
2.4	2.1	10.5	11.7
2.4	1.6	10.6	8.2
2.4	2.2	10.8	9.4
2.5	1.1	12.4	12.2
2.6	3.0	13.0	12.2
2.8	3.1	13.0	8.9
2.9	3.4	13.6	11.1
3.0	1.8	13.8	15.2
3.2	3.2	15.4	18.8
3.3	3.3	16.8	15.8
3.4	1.8	19.3	17.6
3.4	3.4	20.0	15.8
3.6	3.5	20.5	28.8
3.6	2.2	21.8	18.1
3.6	2.8	22.0	25.5
3.9	2.2	23.5	16.4
4.4	4.4	24.9	16.4
4.6	4.8	24.9	20.4
4.6	3.3	26.0	32.4
5.4	4.6	30.0	33.4
5.6	4.3	31.0	41.9
5.7	9.0	33.5	32.9
5.8	3.7	34.5	27.0
5.9	2.9	36.5	37.4
6.0	9.0	37.0	32.9
6.1	5.9	38.0	48.6
6.1	6.1	38.0	40.7
6.2	5.2	67.0	50.9
6.2	10.6		

of interest, hormone assay data. The data are the results of two assay methods for a hormone; the scale of the data as presented is not particularly meaningful, and the original source of the data refused permission to divulge further information. As in a similar example of Leurgans (1980), the old or reference method is being used to predict the new or test method. The overall goal is to see whether we can reproduce the test-method measurements with the reference-method measurements.

In some fields, the quality of the test method has been examined by *ad hoc* rules, such as 'the true intercept should be less than 0.05 in absolute value', or 'the true slope should be at least 0.95', or even 'the true correlation between the measurements should be at least 0.95'. These are usually assessed by hypothesis tests. Alternatively, one might examine prediction and calibration intervals over a range of interest and determine whether they are sufficiently 'precise'.

The standard analysis can be attacked on a number of fronts, but this hardly seems the right forum to do so in detail. The usual analysis

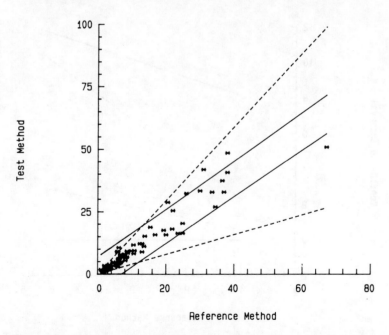

Figure 2.16   *Hormone assay: 95% confidence interval: unweighted, full line: weighted, broken line.*

has the advantage of being almost a protocol, forcing experimenters into producing at least somewhat comparable studies.

The reported results for each method are not exact, i.e., there is some measurement error. It would be interesting to treat this example as a heteroscedastic errors-in-variables problem. See Fuller (1987) for some work done in special cases of the heteroscedastic model.

The data are heteroscedastic (see Figure 2.16). A power-of-the-mean model (2.5) seems reasonable as a model for the variability. Choosing $\theta \simeq 0.90$ seems fair, although the constant-coefficient-of-variation model $\theta = 1.00$ also fits and will be used here. The last data point is outlying in factor space, as the value of the reference method is more than twice as large as any other in the data. With a constant-coefficient-of-variation model, this last point is receiving so little weight as to be almost excluded from the analysis. One difficulty is that the lowest value of the reference method has the largest absolute residual. In the constant-coefficient-of-variation model, this produces a potentially influential outlier (see Chapter 6). We do not pursue this

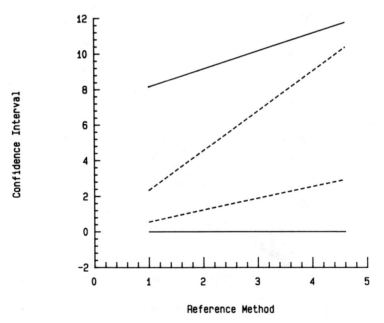

Figure 2.17   *Hormone assay: calibration intervals; unweighted, full line; weighted, broken line.*

in our numerical illustration, but the issue is clearly important and we have seen a similar phenomenon in many other examples. Unusual behavior at the point of smallest predicted variability can indicate failure of the model for the mean at these extreme design points, or be a sign that the variances might be more properly modeled as

$$\sigma_i^2 = \theta_0 + \theta_1 [f(x_i, \beta)]^{\theta_2}$$

This last model reflects the possibility of an extra constant source of variability in the experiment, which is negligible for larger values of the mean but begins to dominate when the mean is small. Alternatively, one could fit the variances as quadratic in the mean response.

A second interesting point here is that a normal probability plot of the studentized residuals suggests a fair amount of right-skewness. An alternative analysis making use of transformations would be interesting.

In Figure 2.16 we plot the 95% prediction interval and in Figure 2.17 the calibration confidence intervals, the latter for estimated reference values less than 5.0. Again, not weighting is seen to be particularly conservative in this range of the data.

In summary, prediction and calibration confidence intervals can be easily constructed for weighted analyses. The effect of ignoring heterogeneity can be quite substantial here, usually more dramatic than the effect on the parameter estimates themselves.

CHAPTER 3

Estimation and inference for variance functions

3.1 Introduction

Variance-function estimation is a form of regression, but one that does not have the wealth of experience and theory behind it as does regression on means. Regression analysis is usually understood to be the examination of the structure of the means as a function of predictors. While there are many components of regression analysis for the means, three stand out as basic. The first is the scatterplot, the standard graphical technique. The second is the existence of a canonical estimation method, least squares. For all its faults, such as sensitivity to outliers and leverage, least squares is a well understood fitting method used as the basis of almost all regression analyses. The third component of regression on means is an off-the-shelf model, the linear model. Even if the mean structure is poorly understood, we can begin an analysis with this standard model, and then try to modify it to account for unexplained effects.

In variance-function estimation, we try to understand the structure of the variances as a function of predictors. It is useful to describe by analogy with regression on means the basic components of variance-function regression. A brief summary of the methods discussed in this chapter is given in Tables 3.1–3.3. Just as for regression on means, there are plotting techniques useful for understanding simple structure. Some of these techniques have been reviewed in the previous chapter. Carrying along the analogy, there are canonical fitting methods, which we describe later in this section under the names pseudo-likelihood and restricted likelihood. The basic idea is first to eliminate the location effects by forming the residuals after an appropriate fit to the mean. One then computes the normal-theory maximum-likelihood estimate of the variance function assuming that

Table 3.1   *Description of methods of variance-function estimation*

Maximum likelihood	Normal-theory maximum likelihood in β and θ
Pseudo-likelihood	Normal-theory MLE when β is set to the current value. When iterated, this is the ordinary MLE if the variance does not depend on the mean
Weighted squared residuals	Regress squared residuals on the variance function, weight inversely with squared current variance estimate
Weighted absolute residuals	Regress absolute residuals on the standard deviation function, weight inverse of current variance estimate
Logarithm method	Regress log absolute residual on log of standard deviation function. Be wary of near-zero residuals
REML	Pseudo-likelihood corrected for leverage. Maximizes marginal posterior for noninformative prior

All the preceding except REML have analogs formed by replacing residuals by sample standard deviations. The following are based on the mean function or x being unknown, and are often used in assays:

Rodbard	Function of mean model. Regress log sample standard deviation on log sample mean
Raab	Function of mean model. Modified functional maximum likelihood with mean function completely unknown
Sadler–Smith	Same as Raab, but means are estimated by sample means

the residuals are the responses and the means are all zero. Just as there are many refinements to least squares which are designed to cope with specific features of data, so too, are there refinements to pseudo-likelihood; some of these are described in the next section. Restricted likelihood is a modification of pseudo-likelihood that corrects for degrees of freedom. The proper analogy with regression on means is

Table 3.2 *Formulae for variance-function estimation: nonregression-based methods; mean $= f(x, \beta)$; standard deviation $= \sigma g(f(x, \beta), z, \theta)$*

Maximum likelihood	Maximize in (β, θ) the function

$$L(\beta, \theta) = -\sum_{i=1}^{N} \log[g(f(x_i, \beta), z_i, \theta)]$$

$$-(N/2)\log\left(N^{-1}\sum_{i=1}^{N}[y_i - f(x_i, \beta)]^2/g^2(f(x_i, \beta), z_i, \theta)\right)$$

Pseudo-likelihood	Fix β at current value β. Maximize in θ the function $L(\beta, \theta)$
Raab	Let $m_i =$ number of replicates at x_i. Maximize in $(\theta, \sigma, \mu_1, \ldots, \mu_N)$ the function

$$L_*(\theta, \sigma, \mu_1, \ldots, \mu_N) = -\sum_{i=1}^{N}(m_i - 1)\log(\sigma \mu_i^{\theta})$$

$$-\tfrac{1}{2}\sigma^{-2}\sum_{i=1}^{N}\sum_{j=1}^{m_i}[(y_{i,j} - \mu_i)/\mu_i^{\theta}]^2$$

Sadler–Smith	Let $\bar{y}_i =$ sample mean at x_i. Maximize in (θ, σ) the function $L_*(\theta, \sigma, \bar{y}_1, \ldots, \bar{y}_N)$

least clear when we try to describe off-the-shelf models. We want simple models relating predictors, which might include the means, to the variances. The first possibility is to model the logarithm of the variances as linear in predictors, e.g.

$$\log(\sigma) = \theta_0 + \theta_1 x + \theta_2 x^{-1}$$

or

$$\log(\sigma) = \theta_0 + \theta_1 \mu(\beta)$$

or

$$\log(\sigma) = \theta_0 + \theta_1 \log[\mu(\beta)]$$

the last being the power-of-the-mean model. Examples of this idea are numerous, and include Box and Hill (1974), Harvey (1976), Just and Pope (1978), Carroll and Ruppert (1982b), etc. The advantage of thinking of the logarithm of the variances as linear in predictors is that the estimated variances are then guaranteed to be positive. Most alternatives to loglinear variance models do not guarantee positive estimated variances. For example, one might hypothesize that the standard deviations or the variances follow a linear model in predictors. The former approach, which we prefer slightly to the latter, has been taken by Glejser (1969), while the latter has been studied by a host of authors, including Hildreth and Houck (1968), Goldfeld and Quandt (1972), and Amemiya (1977). Another possibility is to model the inverses of the variances or standard deviations as linear in predictors; for normally distributed data with known means, this is a canonical form for a generalized linear model.

There are other natural models for the variances. In some instances, the variance might consist of two components, one constant and one depending on the mean. This might suggest a slightly expanded version of the power-of-the-mean model such as

$$\sigma^2 g^2(\mu_i(\beta), z_i, \theta) = \theta_1 + \theta_2 [\mu_i(\beta)]^{\theta_3}$$

Alternatively, one will often see the standard deviation modeled empirically as a quadratic function of the predictors, so that

$$\sigma_i = \sigma g(x_i, \theta) = 1 + \theta_1 x_i + \theta_2 x_i^2$$

If there is more than one predictor, this model can be expanded as in linear regression to include quadratic terms in each variable as well as multiplicative interactions. In addition, one might model the logarithm of the standard deviations as linear or quadratic in the predictors.

Table 3.3 *Formulae for variance-function estimation: regression-based methods; mean = $f(x, \beta)$, standard deviation $\sigma g(f(x, \beta), z, \theta)$, $\hat{\beta}_*$ = current estimate of β, h_{ii} = current leverages, s_i = sample standard deviation at x_i, deviation from fit = $e_i = y_i - f(x_i, \hat{\beta}_*)$*

Method	'Responses'	'Mean function'	'Variance function'
Weighted squared residuals	e_i^2	$\sigma^2 g^2(\mu_i(\hat{\beta}_*), z_i, \theta)$	$g^4(\mu_i(\hat{\beta}_*), z_i, \theta)$
Weighted absolute residuals	$\lvert e_i \rvert$	$\sigma g(\mu_i(\hat{\beta}_*), z_i, \theta)$	$g^2(\mu_i(\hat{\beta}_*), z_i, \theta)$
Logarithm method	$\log \lvert e_i \rvert$	$a + \log [g(\mu_i(\hat{\beta}_*), z_i, \theta)]$	1.0
REML	e_i^2	$\sigma^2 (1 - h_{ii}) g^2(\mu_i(\hat{\beta}_*), z_i, \theta)$	$(1 - h_{ii}) g^4(\mu_i(\hat{\beta}_*), z_i, \theta)$
Rodbard	$\log (s_i)$	$a + \log [g(\bar{y}_i, z_i, \theta)]$	1.0

Another well known variance-function model arises through the random-coefficient linear regression, first studied in detail by Hildreth and Houck (1968). In this model, each individual response has its own regression line given by

$$\sum_{j=1}^{p} \alpha_{i,j} x_i^{(j)}$$

The parameters $\{\alpha_{i,1}, \ldots, \alpha_{i,p}\}$ are most often treated as if they were independently distributed with means $\{\beta_i, \ldots, \beta_p\}$ and unknown variances, so that unconditionally we have the model

$$f(x_i, \beta) = \sum_{j=1}^{p} \beta_j x_i^{(j)}$$

$$\sigma^2 g^2(x_i, \theta) = \sum_{j=1}^{p} \theta_j (x_i^{(j)})^2 \qquad (3.1)$$

In the random-coefficient model, the variance parameters $\{\theta_1, \ldots, \theta_p\}$ are sometimes of interest themselves.

The intent of this chapter is to study in a unified way the estimation of variance-function parameters. The literature tends to treat each variance model as a special case with its own estimation method, usually based on linear regression of functions of absolute residuals against predictors. While this may be useful computationally, it is much less so when trying to make any sense of the area.

There are many instances where the variance function is an important component of independent interest and not just an adjunct for estimating the means. In section 2.9 we introduced an example of method comparison. In our discussions with the engineers who actually do these analyses, it is pretty clear that their interest lies not just in the form of the least-squares line but also, and perhaps even more importantly, in the raw and percentage error one makes in predicting a new observation. The last problem is simply a question of the form and estimation of the variance function.

In section 2.9, we introduced the idea of calibration. Dr C.H. Spiegelman has told us of many instances at the National Bureau of Standards where not only does one want to estimate a new x_0 from an observed y_0, but one also wants a confidence or uncertainty interval. The point estimate is based on the mean function, but the interval is primarily determined by the variance function. For further discussion, see Watters *et al.* (1987).

Here is another application that was brought to our attention by Perry Haaland. In testing for the stability of a biological product, one ordinarily runs an accelerated degradation experiment (Kirkwood, 1977, 1984; Nelson, 1983; Tydeman and Kirkwood, 1984). What follows is a description of the experiments we have seen or read about, and may not be representative of the experience of others. The ultimate goal is to understand the nature of the degradation of the product at a fixed temperature T_0, e.g., 20 °C. Running an experiment at this temperature is impossible because of time constraints; we might want to know the amount of degradation at two years, but for marketing reasons we would like an answer in 60 days. Unfortunately, at 60 days no degradation occurs at T_0 and so no reliable inference can be made. What is done is to run an experiment at elevated temperatures, where the product experiences some significant degradation. Both the mean and the variance of the degradation depend on the temperature at test as well as the time of measurement since initiation of the study. There is a fixed function for the mean, usually an Arrhenius model, and the variances at the test time and temperatures are usually estimated by a replication. Since the variance has not been modeled, we can only estimate the mean degradation at the standard temperature T_0. If the variance were modeled and the model fit to data, we could then undertake an analysis of the distribution of the degradation, e.g., percentiles.

Another area in which variance-function estimation might have important application is off-line quality control (see Taguchi and Wu, 1980; Barker, 1984; Box, 1984, 1987; Box and Meyer, 1985a, b; Leon *et al.*, 1987). As stated by Box and Meyer (1985b), 'one distinctive feature of Japanese quality control improvement techniques is the use of statistical experimental design to study the effect of a number of factors on variance as well as the mean'. One goal of quality-control methods is or should be to design a product that is insensitive to slight changes in either its components or the environment. It makes little sense to design a product that meets specification on average when, with essentially the same components, one can meet specification but have much smaller variability. Thus, we might think of an overall goal as seeking conditions that have a desired mean value for a quality characteristic while at the same time achieving the smallest possible variance. The key point here is that emphasis is no longer only on the mean of the response but also on its variability. As in Box and Meyer (1985b), one might first want to run a screening experiment to understand which factors affect the mean and variance. Based on the screening results, both components can be modeled to solve the optimality problem of minimizing variance subject to a target for the mean. A less direct approach is taken by Taguchi and Wu (1980), who propose to minimize the coefficient of variation as a function of x; their method is closely related to finding the conditions x that minimize the variance of the logarithm of the response. A serious question is how best to design an experiment so as to estimate the variance function efficiently. We will not address this point, which is a subject of ongoing research, but rather we focus only on the estimation of a variance function given the design. Box and Meyer (1985b) have a wonderful discussion notable for its clarity as well as the idea that variance functions can be postulated after data analysis and then fit. This is not so very far from what is usually done for regression on means.

A final application is the estimation of the sensitivity of a chemical or biochemical assay. As discussed by Oppenheimer et al. (1983) and Carroll et al. (1986a), the minimum detectable concentration and the lowest limit of reliable measurement are to a great extent determined by the variance function.

Before we describe methods for variance-function estimation, the generalized least-squares algorithm first needs to be updated to allow for estimating θ. The algorithm as presented here allows a rather general form for the method of estimating θ through solving the equation (3.2). In the next two sections we will discuss some specific algorithms that have been used to estimate the variance parameter.

Algorithm for generalized least squares

Step 1   Start with a preliminary estimator $\hat{\beta}_*$.

Step 2   Estimate the parameter θ by solving an equation of the form

$$0 = \sum_{i=1}^{N} H(y_i, z_i, \mu_i(\hat{\beta}_*), \theta) \qquad (3.2)$$

Compute estimated weights

$$\hat{w}_i = 1.0/g^2(\mu_i(\hat{\beta}_*), z_i, \hat{\theta}) \qquad (3.3)$$

Step 3   Let $\hat{\beta}_G$ be the weighted least-squares estimate using the estimated weights (3.3).

Step 4   Update the preliminary estimator by setting $\hat{\beta}_* = \hat{\beta}_G$, and the estimate of θ and the weights as in (3.2) and (3.3).

Step 5   Repeat steps 3 and 4 $\mathbb{C} - 1$ more times, where $\mathbb{C}$ is the number of cycles in generalized least squares and is chosen by the experimenter. Alternatively, one could stop when there is little change in β and θ. See section 3.2 for a discussion of this issue.

Step 6   Full iteration corresponds to setting the number of cycles $\mathbb{C} = \infty$. This is the same as solving the equations

$$0 = \sum_{i=1}^{N} f_\beta(x_i, \beta)[y_i - f(x_i, \beta)]/g^2(\mu_i(\beta), z_i, \theta)$$

$$0 = \sum_{i=1}^{N} H(y_i, z_i, \mu_i(\beta), \theta) \tag{3.4}$$

which will usually be done by a separate method, since iterative weighted least squares may not always converge. *End*

The asymptotic theory for $\hat{\beta}_G$ discussed in section 2.2 is unchanged by estimating θ. Essentially independent of the method of estimating θ or the number of cycles $\mathbb{C}$, the generalized least squares estimate has the same asymptotic distribution as the weighted least-squares estimate with known weights, i.e., it is normally distributed with mean β and covariance matrix $(\sigma^2/N)S_G^{-1}$, where S_G is defined in (2.4).

When one must estimate θ, the asymptotic theory, which gives the same standard errors for $\hat{\beta}$ as if θ were known, generally is optimistic. While this fact is not the primary or even a particularly strong justification for studying variance-function estimation, it does give an added incentive. Williams (1975) states that 'both analytic and empirical studies of a variety of linear models indicate that...the ordering by efficiency of (estimates of β)...in small samples is in accordance with the ordering by efficiency' of estimates of θ. There is a particularly interesting Monte Carlo simulation in the book by Goldfeld and Quandt (1972) which illustrates this phenomenon. Theoretical work by Toyooka (1982) and Rothenberg (1984) shows that, for normally distributed data when the variances depend on predictors but not the mean, the covariance matrix of $N^{1/2}(\hat{\beta}_G - \beta)$ is given by

$$\sum_{N}(\mathbb{C}, \hat{\theta}) \simeq \sigma^2 S_G^{-1} + N^{-1} V(\hat{\theta}) \qquad (3.5)$$

These authors show that the correction term $V(\hat{\theta})$ is an increasing function of the asymptotic variance of $\hat{\theta}$. Thus, for normally distributed data the better one estimates θ the better one will estimate β. This result is sketched in Chapter 7. Rothenberg's proof makes clever use of complete sufficiency (see also Eaton, 1985).

3.2 Pseudo-likelihood estimation of variance functions

In this section we will discuss a method of estimation called pseudo-likelihood. If the variance does not depend on the mean, pseudo-likelihood includes the usual normal-theory maximum-likelihood estimate as a special case.

We will use the term pseudo-likelihood to denote the following analog to generalized least squares. Like generalized least squares, this method makes no distributional assumptions and instead relies only on the basic mean and variance model (2.1), but like generalized least squares its efficiency can be seriously diminished by slight deviations from normality. Pseudo-likelihood estimates of θ are based on pretending that the regression parameter β is known and equal to the current estimate $\hat{\beta}_*$, and then estimating θ by maximum likelihood assuming normality, i.e., maximizing in θ and σ the loglikelihood $L(\hat{\beta}_*, \theta, \sigma)$, where

$$L(\beta, \theta, \sigma) = -N \log(\sigma) - \sum_{i=1}^{N} \log \left[g(\mu_i(\beta), z_i, \theta) \right]$$

$$- (2\sigma^2)^{-1} \sum_{i=1}^{N} \left\{ [y_i - f(x_i, \beta)] / g(\mu_i(\beta), z_i, \theta) \right\}^2$$

The reason that we call this method pseudo-likelihood is that even its fully iterated version is not the same as the actual normal-theory maximum-likelihood estimate of θ unless $\hat{\beta}_*$ is the maximum-likelihood estimate of β. The maximum-likelihood estimate of β is a generalized least-squares estimate only in the case that the variance does not depend on the mean. Write the residuals as

$$r_i(\beta, \theta) = \frac{y_i - f(x_i, \beta)}{g(\mu_i(\beta), z_i, \theta)} \qquad (3.6)$$

Writing

$$\hat{\sigma}^2(\beta, \theta) = N^{-1} \sum_{i=1}^{N} [y_i - f(x_i, \beta)]^2 / g^2(\mu_i(\beta), z_i, \theta) \qquad (3.7)$$

the pseudo-likelihood estimate $\hat{\theta}_{PL}$ maximizes in θ the log-pseudo-likelihood $L_{PL}(\hat{\beta}_*, \theta)$, where

$$L_{PL}(\beta, \theta) = - N \log [\hat{\sigma}(\beta, \theta)] - \sum_{i=1}^{N} \log [g(\mu_i(\beta), z_i, \theta)] \qquad (3.8)$$

Define also

$$v(i, \beta, \theta) = \log [g(\mu_i(\beta), z_i, \theta)] \qquad (3.9)$$

Taking derivatives with respect to θ, $\hat{\theta}_{PL}$ is the solution, assuming it exists, to the equation

$$\sum_{i=1}^{N} [r_i^2(\hat{\beta}_*, \theta) - \hat{\sigma}^2(\hat{\beta}_*, \theta)] v_\theta(i, \hat{\beta}_*, \theta) = 0$$

where v_θ is the derivative of v with respect to θ. This has the form of a set of normal equations, with the 'design' being the terms v_θ. Morton (1987b) has pointed out that in some contexts, in order to avoid extreme leverage effects, it can sometimes be better to replace v_θ by other terms.

The pseudo-likelihood estimate can be computed by one of three devices. The first, based on generalized least squares applied to squared residuals, is discussed in subsection 3.3.1. A second method is applicable if θ is scalar as in the power-of-the-mean model. In this case one can compute (3.8) along a grid of values in a range of interest and choose the value of θ resulting in a maximum. As a third method, it is possible to use nonlinear least-squares programs for computing $\hat{\theta}_{PL}$.

Let

$$\dot{g}(\beta, \theta) = \left(\prod_{i=1}^{N} g(\mu_i(\beta), z_i, \theta) \right)^{1/N}$$

Set all 'responses' to be zero and the 'regression function' to be

$$D_i(\beta, \theta) = \dot{g}(\hat{\beta}_*, \theta) [y_i - f(x_i, \beta)] / g(\mu_i(\hat{\beta}_*), z_i, \theta)$$

It is easy to show that minimizing the sum of $D_i^2(\beta, \theta)$ is equivalent to maximizing (3.8). When it converges, nonlinear least squares applied to these quantities yields the pseudo-likelihood estimate of θ and a generalized least-squares estimate of β. If the variance function does not depend on the mean, this is one way to compute the maximum-likelihood estimate of β and θ simultaneously.

Under weak moment assumptions and without making any distributional assumptions, the pseudo-likelihood estimate of θ is asymptotically normally distributed with mean θ and a variance depending on θ, σ, and the starting estimate $\hat{\beta}_*$. The general expression for this variance is rather complicated and is given in Davidian and Carroll (1987).

Here is one important point with some practical implications. Suppose that the variance function depends on the mean response. If the errors (2.21) are symmetrically although not necessarily identically distributed, then the variance of $\hat{\theta}_{PL}$ is an increasing function of the variance of the estimate of β. Thus one ought to use at least $\mathbb{C} = 2$ cycles in the generalized least-squares algorithm since the estimate with $\mathbb{C} = 1$ is based on a less efficient estimate of β. This has important consequences for variance-function estimation, suggesting that such estimation should not be based on least-squares residuals. This fact carries over to almost every method of variance-function estimation.

For small σ when the errors (2.21) are identically distributed, we have the approximate covariance

$$\text{Covariance}(N^{1/2}(\hat{\theta}_{PL} - \theta)) \simeq (1/2 + \kappa/4)\zeta(\mu)^{-1}$$

where κ is the kurtosis of the errors and is equal to zero for normally distributed data, and $\zeta(\mu)$ is the sample covariance matrix of the derivatives v_θ of v with respect to θ (see equation (3.9)). The fact that the variance of $\hat{\theta}_{PL}$ depends on the fourth moments of the errors suggests that this estimate will be affected adversely by outliers. In the power-of-the-mean-variance model, $v(i, \beta, \theta)$ is the logarithm of the mean response.

One criticism of pseudo-likelihood is that it takes no account of the loss in degrees of freedom that results from estimating β. We have already adjusted for this partially by using in (2.14) the divisor $(N - p)$ rather than N, the latter defining the pseudo-likelihood estimate. Harville (1977) presents a thorough discussion of this issue with emphasis on variance-components models in the analysis of variance. The effect of applying pseudo-likelihood directly to estimate θ seems to be a bias that depends on the ratio p/N, where p is the number of regression parameters. In designed experiments such as fractional factorials, p is often rather large relative to the sample size N and some adjustment might be necessary.

Suppose that the variance does not depend on the mean, so that

$$\text{Var}(y_i) = \sigma^2 g^2(z_i, \theta)$$

Let $\hat{\beta}$ be a generalized least-squares estimate. Then pseudo-likelihood solves in (θ, σ) the equations

$$\sum_{i=1}^{N} e_i^2 \left(\frac{1}{v_\theta(i, \theta)} \right) = \sum_{i=1}^{N} \left(\frac{1}{v_\theta(i, \theta)} \right) \qquad (3.10)$$

where

$$e_i = e_i(\hat{\beta}, \theta, \sigma) = [y_i - f(x_i, \hat{\beta})]/[\sigma g(z_i, \theta)]$$

As discussed prior to equation (2.25), if $\hat{\beta}$ is a generalized least-squares estimate then at the correct value of θ we can approximate the vector of the N residuals by $(I - H)E$. Here E is a random vector of N elements with mean zero and identity covariance, and H is the $N \times N$ hat matrix $H = H(\beta, \theta) = X_*(X_*^T X_*)^{-1} X_*^T$, where $X_* = X_*(\beta, \theta)$ is the $N \times p$ matrix with ith row the transpose of the column vector

$$f_\beta(x_i, \hat{\beta})/g(z_i, \theta) \qquad (3.11)$$

The leverage values h_{ii} are the diagonals of the hat matrix H. Note that after making this approximation e_i^2 has mean $(1 - h_{ii})$. Also, since H is an idempotent matrix of rank p, the sum of the leverage values is p. We thus find that the left-hand side of (3.10) has approximate expectation given by

$$\left[\frac{N - p}{\sum_{i=1}^{N} \{v_\theta(i, \theta)[1 - h_{ii}(\beta, \theta)]\}} \right]$$

To account for the loss in degrees of freedom due to estimating β, the suggestion is to equate the left-hand side of (3.10) to its approximate expectation, thus solving in θ and σ the equations

$$\sum_{i=1}^{N} e_i^2(\hat{\beta}, \theta, \sigma) \left[\frac{1}{v_\theta(i, \theta)} \right] = \left[\frac{N - p}{\sum_{i=1}^{N} \{v_\theta(i, \theta)[1 - h_{ii}(\hat{\beta}, \theta)]\}} \right] \qquad (3.12)$$

This argument works even when the variance function depends on β, as long as σ is small, the analysis being based on (2.31).

An alternative device to account for the loss of degrees of freedom relies on Bayesian ideas. See Patterson and Thompson (1971) and Harville (1977) for the standard method, called restricted maximum likelihood and often abbreviated by REML. When the variances depend on the mean, this might more properly be called restricted pseudo-likelihood, but we will use Harville's terminology. A number of alternatives are discussed by Harville (1977) (see also Green, 1985). Assume that the variances do not depend on the mean. For this case, maximum-likelihood estimation of β and θ belongs to the class of pseudo-likelihood estimates of θ combined with a generalized least-squares estimate of β. Pretend that the parameters (β, θ, σ) have an improper prior density $\pi(\beta, \theta, \sigma) \equiv \sigma^{-1}$. Either proper or other improper priors could be used to generate alternative estimators, at least in principle. The posterior density of (β, θ, σ) given the data is proportional to the likelihood of the data. Harville (1977) argues that estimates of θ should be the mode of appropriate posterior distributions. For example, the maximum-likelihood estimate of θ can be obtained as the θ component to the mode of the joint posterior density for (β, θ, σ).

Write the weights as $w_i(\theta) = 1/g_i^2(\theta)$ and note that the likelihood is proportional to

$$p(\beta, \theta, \sigma) = \left(\prod_{i=1}^{N} w_i(\theta) \right)^{1/2} \sigma^{-N}$$

$$\times \exp\left(-(2\sigma^2)^{-1} \sum_{i=1}^{N} w_i(\theta)[y_i - f(x_i, \beta)]^2 \right)$$

Thus, the marginal posterior density of (θ, σ) is proportional to

$$p(\theta, \sigma) = \sigma^{-1} \int p(\beta, \theta, \sigma) \, d\beta \tag{3.13}$$

The integral is generally hard to compute in closed form for a nonlinear regression model, so it can be difficult to compute numerically because of the need to compute multiple integrals. Following Box and Hill (1974) and Beal and Sheiner (1985), note that if $\hat{\beta}$ is a generalized least-squares estimate for β, then we have the linear approximation

$$f(x_i, \beta) \simeq f(x_i, \hat{\beta}) + u_i^T \gamma$$

where

$$u_i = f_\beta(x_i, \hat{\beta}) \qquad \text{and} \qquad \gamma = \beta - \hat{\beta}$$

Replacing $f(x_i, \beta)$ in (3.13) by its linear expansion and integrating with respect to γ, the marginal posterior can now be computed exactly as proportional to

$$p_A(\theta, \sigma) = \frac{[\prod w_i(\theta)]^{1/2} \exp\{-[(N-p)/2]\hat{\sigma}_G^2(\theta)/\sigma^2\}}{\sigma^{N-p} \mathrm{Det}^{1/2}[S_G(\theta)]} \qquad (3.14)$$

where $\hat{\sigma}_G(\theta)$ and $S_G(\theta)$ are given by (2.14) and (2.15) and Det stands for the determinant of the matrix. Integrating with respect to σ, the restricted maximum-likelihood estimate for θ is seen to maximize in θ the expression

$$\{[\prod w_i(\theta)] \mathrm{Det}[\hat{\sigma}_G^2(\theta)S_G^{-1}(\theta)]/[\hat{\sigma}_G^2(\theta)]^N\}^{1/2} \qquad (3.15)$$

It is interesting to compare this to the pseudo-likelihood estimate which maximizes

$$\{[\prod w_i(\theta)]/[\hat{\sigma}_G^2(\theta)]^N\}^{1/2} \qquad (3.16)$$

If the variances depend on the regression parameter β, then the Bayesian arguments can be extended by replacing $w_i(\theta)$ by $w_i(\theta, \hat{\beta})$. This question has not been explored in great detail. Box and Hill (1974) and Beal and Sheiner (1985) base their calculations on an iterative process much like our algorithm for generalized least squares, i.e., one starts with an estimate of β, then one estimates θ by restricted maximum likelihood followed by an updating of β, etc.

By using matrix derivatives as in Nel (1980), we show in Chapter 7 that the posterior mode of (3.15) is exactly the solution to (3.12) based on equating sums of squares to their expectations.

Pseudo-likelihood and restricted likelihood are the standard methods of variance estimation. They are widely applicable, require almost no distributional assumptions, and generalize rather easily in principle to problems with a nondiagonal covariance, such as variance-components models. They can be very sensitive to outliers because their influence functions are quadratic in the errors (see Chapter 6). For these and other reasons, a variety of alternative methods have been proposed in the literature and are discussed in the next section.

3.3 Other methods of variance-function estimation

Pseudo-likelihood and its bias-corrected versions have the advantage of not requiring distributional assumptions and are in this sense analogous to generalized least squares. The methods extend easily to the case of nonindependent observations, for which the covariance matrix of the responses will no longer be diagonal. If the observations are independent, there are a variety of techniques that serve as competitors to pseudo-likelihood. In this section we review many of the techniques that have been proposed in the literature.

Alternative methods for estimating θ in the parametric model (2.1) can be classified into four rough categories. The first consists of those which make only the assumption that the mean and variance model (2.1) holds; the representatives of this class studied here include, in addition to pseudo-likelihood, weighted least squares on squared residuals. The second class of methods is based on weighted regression of transformations of absolute residuals on their expected values, assuming that the errors (2.21) are independent and identically distributed. A third way to estimate θ is to assume that the data are generated from a flexible class of distributions indexed by θ and then to apply likelihood techniques. Finally, especially in the context of assays, one might observe replicate responses at each value of the predictor and use transformations of the sample variances to estimate θ. In this section we will list some members of each general class and give a few comparisons among the various methods. Our discussion is based on Davidian and Carroll (1987).

Most methods for variance-function estimation are based on generalized least-squares ideas, with 'responses' being transformations of absolute residuals or sample standard deviations. The analogy with ordinary regression then becomes obvious, and suggests for example that one might apply the range of diagnostics available for ordinary regression, such as residual plots, influence diagnostics, stepwise selection, etc. In Chapter 6 we develop some general diagnostics and apply them to pseudo-likelihood.

It is also useful to remember that if there are constraints on the variance parameter θ, then these constraints should be incorporated into the estimation process. For example, in the power-of-the-mean model it often makes sense to restrict θ to be at least zero but not more than 1.5.

3.3.1 Weighted least squares on squared residuals

The motivating idea for these methods is that the expectation of squared residuals is approximately the variance. Thus, as noted by Hildreth and Houck (1968), Goldfeld and Quandt (1972), and Jobson and Fuller (1980), we can consider a nonlinear regression problem where the 'responses' are squared residuals and the 'regression function' is its approximate expectation $\sigma^2 g^2(\mu(\beta), z, \theta)$. The first suggestion is to estimate θ by minimizing in σ and θ the least-squares equation

$$\sum_{i=1}^{N} \{[y_i - f(x_i, \hat{\beta}_*)]^2 - \sigma^2 g^2(\mu_1(\hat{\beta}_*), z_i, \theta)\}^2 \qquad (3.17)$$

For normal data the squared residuals are themselves heteroscedastic with approximate variance proportional to $\sigma^4 g^4(\mu_i(\beta), z_i, \theta)$, so that one is immediately led to generalized least squares.

The essence of the problem is seen to be an approximate heteroscedastic regression model. Given an estimate $\hat{\beta}$ of β, weighted squared residuals estimates are generalized least-squares estimates obtained from the 'model' with

$$\begin{aligned}
&\text{'Responses'} = [y_i - f(x_i, \hat{\beta})]^2 \\
&\text{'Parameters'} (\sigma, \theta) \\
&\text{'Regression function'} = \sigma^2 g^2(\mu_i(\hat{\beta}), z_i, \theta) \\
&\text{'Variances' proportional to } g^4(\mu_i(\hat{\beta}), z_i, \theta)
\end{aligned} \qquad (3.18)$$

Note that this problem is formally the same as that studied in the previous chapter. It is interesting to ask what happens when we fully iterate generalized least squares, which in the terminology of Chapter 2 is a quasi-likelihood estimate. It can be shown that such an estimate of θ, when the process converges, is exactly the pseudo-likelihood estimate.

Jobson and Fuller (1980) among others have noted that if the unweighted least-squares estimate of β is used in minimizing (3.17), then the squared deviations $[y_i - f(x_i, \hat{\beta}_{LS})]^2$ have approximate expectation different from $\sigma^2 g^2(\mu_i(\hat{\beta}_{LS}), z_i, \theta)$ by a factor depending on the ith leverage value from the unweighted least-squares fit. For a generalized least-squares fit $\hat{\beta}_G$ with ith leverage value $h_{ii}(GLS)$, the squared deviations have approximate expectation

$$\sigma^2 [1 - h_{ii}(GLS)] g^2(\mu_i(\beta), z_i, \theta)$$

For a general sequence of constants c_i, one might consider updating θ from a previous estimate $\hat{\theta}_{WLS}$ by minimizing in (σ, θ)

$$\sum_{i=1}^{N} \frac{\{[y_i - f(x_i, \hat{\beta}_G)]^2 - \sigma^2 [1 - h_{ii}(GLS)] g^2(\mu_i(\hat{\beta}_G), z_i, \theta)\}^2}{c_i g^4(\mu_i(\hat{\beta}_G), z_i, \hat{\theta}_{WLS})}$$

Choosing $c_i = 1 - h_{ii}(\text{GLS})$, fully iterating yields the restricted maximum-likelihood estimate. It is interesting to note that this is neither the natural nor obvious choice for c_i. Assuming normality, the squared deviations have approximate variance proportional to

$$[1 - h_{ii}(\text{GLS})]^2 g^4(\mu_i(\beta), z_i, \theta)$$

which suggests $c_i = [1 - h_{ii}(\text{GLS})]^2$. With this last choice, the updated estimate would solve the pseudo-likelihood equation (3.10) except that the standardized residual e_i defined just after (3.10) would be replaced by a studentized residual as defined in section 2.7. We do not know if such a method has been studied in the literature, but it can be rather easy to compute, and we have used it with some success.

For symmetrically distributed errors or as an approximation for small σ, the weighted squared residual estimate $\hat{\theta}_{\text{WLS}}$ is known to be asymptotically more efficient than the unweighted estimator minimizing (3.17) (see Davidian and Carroll, 1987). Under very general conditions, they show that any weighted estimator based on (3.18) has the same asymptotic distribution as the pseudo-likelihood estimator $\hat{\theta}_{\text{PL}}$.

3.3.2 Methods based on transformations of absolute residuals

Since least-squares operations are adversely affected by outlying responses (Huber, 1981), using squared residuals is a method particularly prone to degraded performance due to unexpectedly large residuals which, when squared, become very outlying. Cohen et al. (1984) note this and suggest that better performance can be obtained by using absolute residuals (see also Glejser, 1969; Harvey, 1976; Judge et al., 1985). Harvey (1976) has suggested regressing the logarithm of the absolute residuals on the logarithm of the standard-deviation function. These methods, which we will discuss in detail below, are examples of techniques based on regressing transformations of absolute residuals on their expected values. One such class of transformations is the power class

$$h(v, \lambda) = \begin{cases} v^\lambda & \text{for } \lambda \neq 0 \\ \log(v) & \text{for } \lambda = 0 \end{cases}$$

If we write the absolute deviation as

$$d_i(\beta) = |y_i - f(x_i, \beta)|$$

then assuming the errors (2.21) are independent and identically distributed we have that, for a constant c_0,

$$E(h(d_i(\beta), \lambda)) = \begin{cases} c_0 g^\lambda(\mu_i(\beta), z_i, \theta) & \text{for } \lambda \neq 0 \\ c_0 + \log[g(\mu_i(\beta), z_i, \theta)] & \text{for } \lambda = 0 \end{cases}$$

(3.19)

Further, for a constant c_1 the variance of the transformed absolute deviations satisfies

$$\text{Variance}(h(d_i(\beta), \lambda)) = \begin{cases} c_1 g^{2\lambda}(\mu_i(\beta), z_i, \theta) & \text{for } \lambda \neq 0 \\ c_1 & \text{for } \lambda = 0 \end{cases}$$

(3.20)

What this means is that using weighted regression, one can calculate an estimate of θ based on any power of the absolute deviations from a fitted line. The appropriate 'regression function' is given by (3.19), while the 'variance function' is given by (3.20). Choosing $\lambda = 2$ yields the previously discussed squared residual methods, $\lambda = 1$ the absolute residuals, and $\lambda = 0$ their logarithms.

3.3.3 Methods based on absolute residuals

It has been suggested that using absolute rather than squared residuals in least-squares operations will be more efficient because the former will not have as wild outliers as the latter. Although we will not pursue this idea, at least for normally distributed data it might be appropriate to account for leverage as we have done in section 3.1 and subsection 3.3.1.

The basic requirement is a moment assumption of a different form, namely that the absolute deviation has expected value

$$E|y_i - f(x_i, \beta)| = c_f g(\mu_i(\beta), z_i, \theta)$$

(3.21)

Here, the constant c_f is a constant of proportionality which is independent of the design point x_i. For most purposes, if the basic mean and variance assumption (2.1) and (3.21) hold simultaneously, then the errors (2.21) are independent and identically distributed. Of course, one might be willing to assume (3.21) rather than (2.1), and if one assumes as well that $f(x_i, \beta)$ is the median of y_i rather than the mean, then one would be led to weighted least-absolute-values regression. Many people have suggested that, for very skewed data, the median may be a more easily interpretable measure than the mean; see Snee (1986) for an example. We do not pursue this idea, and instead stay within the standard framework of a mean and variance model. The assumption (3.21) is violated in the case of the Poisson or gamma distributions, although in both of these cases the assumption is nearly correct if the mean responses are large. Assumption (3.19) is made implicitly when one forms absolute residual plots.

The absolute residuals will have expectation given approximately by (3.19) with variance approximately proportional to g^2. This suggests estimating θ by generalized least-squares techniques applied to the absolute residuals. For a given estimate $\hat{\beta}$, the calculations are based on the 'model' with

$$
\begin{aligned}
&\text{'Responses'} = |y_i - f(x_i, \hat{\beta})| \\
&\text{'Parameters'} = (c_f, \theta) \\
&\text{'Regression function'} = c_f g(\mu_i(\hat{\beta}), z_i, \theta) \\
&\text{'Variance function' proportional to } g^2(\mu_i(\hat{\beta}), z_i, \theta)
\end{aligned}
\tag{3.22}
$$

Having obtained such an estimate of θ, we then update the estimate of β by generalized least squares. This process can be repeated a few times, getting a new estimate of θ, then of β, etc. Convergence is not guaranteed, but as mentioned in section 3.1 for symmetrically distributed errors the iterated estimate of θ will be more efficient. In our experience with the power-of-the-mean model (2.5), the estimate of θ based on unweighted least-squares residuals is almost always too small, so that some iteration is useful.

Davidian and Carroll (1987) compare the efficiency of weighted squared residual or pseudo-likelihood estimation of θ with weighted absolute residual estimation. If the variance function does not involve the mean and either (1) the errors are symmetric or (2) σ is small, they find that

$$
\left.\begin{aligned}
&\text{Asymptotic relative efficiency} \\
&\text{of absolute residual estimation}
\end{aligned}\right\} = \frac{(2 + \kappa)(1 - \delta)}{4\delta}
\tag{3.23}
$$

where κ is the kurtosis of the errors and δ is the variance of the absolute errors (2.21). Equation (3.23) is also the asymptotic relative efficiency of the mean absolute deviation with respect to the sample variance in a one-sample problem. For normally distributed data, using absolute residuals results in a 12% loss of efficiency. For double-exponential errors with the density $\frac{1}{2}\exp(-|x|)$, there is a 25% gain in

Table 3.4 *Efficiencies at the contaminated normal distribution of two estimates of the variance function: weighted absolute residual estimation and pseudo-likelihood or weighted squared residual estimation*

Contamination fraction	Gain in efficiency (%) by using absolute residuals
0	− 12.4
0.001	− 5.2
0.002	1.6
0.01	43.9
0.05	103.5

efficiency by using absolute residuals. Huber (1981, p. 3) presents an interesting comparison for the class of contaminated normal distributions, in which a large fraction $(1 - \alpha)$ of the data have a normal distribution and a fraction α are contaminated in that they have a normal distribution with the same mean but three times larger standard deviation. The gain in efficiency for using absolute residuals is given in Table 3.4, with negative values indicating a loss. Huber calls these numbers 'disquieting', which indeed they are because just two 'bad' observations in 1000 suffice to offset the superiority of squared residuals over absolute residuals when estimating the variance function. These numbers hint at the lack of robustness of the standard procedures, an issue that is discussed at length in Chapter 6. In the terminology of that chapter, the influence function of pseudo-likelihood is quadratic in the errors, while that of weighted absolute residuals is linear in the absolute errors. While neither method is robust, the latter is relatively more so.

3.3.4 Methods based on the logarithm of absolute residuals

Harvey (1976) has suggested regressing the logarithm of the absolute residuals on the logarithm of their approximate expected values, i.e., $\log [\sigma g(\mu_i(\hat{\beta}_*), z_i, \theta)]$. Assuming that the errors are independent and identically distributed, this should be approximately a homoscedastic regression. The calculations are easy since only an ordinary nonlinear least-squares program is required, although a practical problem can arise if one of the residuals is very near zero, in which case taking logarithms induces a rather large and artificial outlier. To avoid this potential difficulty with 'inliers', for fitting the variance model we suggest deleting a few of the smallest absolute residuals. The *ad hoc* nature of this deletion, as well as the lack of clarity in choosing how many points to delete, points out a need for a more complete study of regression diagnostics and robust estimators (see Chapter 6). The problem with small absolute residuals necessitates extreme care.

Iteration is useful here. One starts with the unweighted least-squares estimate of β, obtains an estimate of θ, then a quasi-likelihood generalized least-squares estimate of β, then an updated estimate of θ, etc. Convergence is not guaranteed. At least for the power-of-the-mean model, we have found that using unweighted least-squares residuals in the estimation of θ usually produces a noticeable underestimate. For symmetric data, not iterating is inefficient.

If the variances do not depend on the mean and either (1) the errors are symmetric or (2) σ is small, then the effect of estimating β can be ignored, and the logarithm method has asymptotic efficiency

$$(2 + \kappa)/\mathrm{Variance}[\log(\varepsilon^2)]$$

with respect to pseudo-likelihood. For normally distributed data, this efficiency is 40.5%. For contaminated normal data, in Table 3.5 we list

Table 3.5 *Efficiencies at the contaminated normal distribution of two estimates of the variance function with respect to pseudo-likelihood: weighted absolute residual estimation and the logarithm method*

Contamination fraction	Gain in efficiency (%) by using absolute residuals	Gain in efficiency (%) by Harvey's method
0	− 12.4	− 59.5
0.001	− 5.2	− 56.0
0.002	1.6	− 52.0
0.01	43.9	− 28.0
0.05	103.5	22.0

the efficiency of the logarithm method as well as that based on weighted absolute residuals. Certainly, the gain in efficiency with respect to pseudo-likelihood is noticeable at 5% contamination. A cousin to the logarithm method, explained later in this section, is in common use in radioimmunoassay.

3.3.5   Assuming a flexible class of distributions

One can argue that pseudo-likelihood fits into this scheme. Essentially, pseudo-likelihood assumes that the data are normally distributed with known mean β, and then one applies maximum likelihood to estimate θ; β is estimated by generalized least squares. Luckily, this scheme is a method of moments so that pseudo-likelihood estimation of θ is consistent and asymptotically normal under very general conditions, without assuming normality. This perspective suggests that one might define a different class of distributions for the error term. Such a class might include skewed distributions such as the gamma or Poisson, as well as the normal distributions. This is one way to motivate the idea of extended quasi-likelihood due to Nelder and Pregibon (1987) and discussed in McCullagh and Nelder (1983), as well as a similar family recently proposed by Efron (1986). We will discuss only the former suggestion. One point to keep in mind concerns distributional differences. As we vary θ, two components of the problem vary. First, of course, is the structure of the variances, which at the initial stage of analysis might be of most importance. The second component is the distribution associated with the value of θ. If estimation and inference suggest major differences between an estimate of θ and a hypothesized value θ_0, such indication might well reflect differences in distribution rather than substantial doubt about the variance function implied by θ_0.

Extended quasi-likelihood   Nelder and Pregibon (1987) attempt to define a family of distributions indexed by a parameter θ which has in addition the following properties: (i) has mean and variance functions given by (2.1); (ii) has maximum-likelihood estimate given by generalized least squares; and (iii) includes homoscedastic normal, gamma, and inverse Gaussian distributions as special cases in the power-of-the-mean model (2.18). It is not true that all three can be achieved simultaneously (Bar-Lev and Ennis, 1986), but at least for small σ all three can be achieved approximately. As Morton (1987a, b) and Davidian and Carroll (1988) have pointed out, Nelder and Pregibon's estimate of θ is generally inconsistent (see the end of this subsection).

Consider an exponential family with log-density given by

$$\log [f(y|\xi, \sigma)] = [y\xi - b(\xi)]/\sigma^2 + c(y; \sigma)$$

for some functions $b(\cdot)$ and $c(\cdot; \cdot)$. Since in an exponential family the mean is given by $\mu = b_\xi(\xi)$ and the variance is given by $\sigma^2 b_{\xi\xi}(\xi)$, we find that in terms of μ and θ

$$\mu = b_\xi(\xi) = [\partial b(\xi)/\partial \mu]/[\partial \xi/\partial \mu]$$
$$g^2(\mu, z, \theta) = \partial^2 b(\xi)/\partial \xi^2 = \partial \mu/\partial \xi = (\partial \xi/\partial \mu)^{-1}$$

This yields

$$b(\mu) = \int^\mu [w/g^2(w, z, \theta)]\, dw$$

$$\xi = \int^\mu [1/g^2(w, z, \theta)]\, dw$$

Hence, we get that

$$l_Q(\mu, y, \theta) = y\xi - b(\xi) = \int^\mu \{(y - w)/\, g^2(w, z, \theta)\}\, dw$$

The function l_Q is the quasi-likelihood kernel. Subtracting the same integrals over the range up to y, we find that for a function c_*, the log density is

$$\ln [f(y|\mu, \theta, \sigma)] = \sigma^{-2} \int_y^\mu [(y - w)/g^2(\mu, z, \theta)]\, dw + c_*(y, \theta, \sigma)$$
$$= \sigma^{-2} [l_Q(\mu, y, \theta) - l_Q(y, y, \theta)] + c_*(y, \theta, \sigma) \qquad (3.24)$$

In general (in order for f to be a density) c_* will depend on μ, in which case the maximum-likelihood estimate of the mean function will not be a generalized least-squares estimate. However, if σ is small we can get an approximation for c_* that does not depend on μ. In this case

$$\int_y^\mu [(y - w)/g^2(\mu, z, \theta)]\, dw \simeq -\tfrac{1}{2}(y - \mu)^2/[\sigma g(y, z, \theta)]^2$$

Substituting this approximation into (3.24), in order for the resultant to be a density, we must have

$$c_*(y, \theta, \sigma) = -\log [(2\pi)^{1/2}\sigma g(y, z, \theta)]$$

Combining these, leads to the extended quasi-loglikelihood given by

$$l_Q^*(\mu, y, \theta, \sigma) = \sigma^{-2}[l_Q(\mu, y, \theta) - l_Q(y, y, \theta)]$$
$$- \log[(2\pi)^{1/2}\sigma g(y, z, \theta)]$$

The extended quasi-loglikelihood for the entire data is

$$\sum_{i=1}^{N} l_Q^*(\mu_i(\beta), y_i, \theta, \sigma) \qquad (3.25)$$

For given θ and σ, β is to be estimated by generalized least squares, while for given β and θ the estimate of σ solves

$$N^{-1} \sum_{i=1}^{N} [l_Q(\mu_i(\beta), y_i, \theta) - l_Q(y_i, y_i, \theta)] = \sigma^2(\beta, \theta)$$

The parameters are to be estimated by jointly maximizing (3.25). For the power-of-the-mean model (2.5), with $\theta = 0$, l_Q^* is the loglikelihood of the normal distribution, while it is the loglikelihood of the inverse Gaussian distribution for $\theta = 1.5$. Otherwise, l_Q^* is not exactly a loglikelihood, which distinguishes extended quasi-likelihood from pseudo-likelihood. For fixed σ, when the mean is large, l_Q^* with $\theta = 1.0$ is approximately the loglikelihood of a gamma distribution, while for $\sigma = 1$, $\theta = 0.5$, and μ large, l_Q^* approximates the Poisson loglikelihood.

The fact that l_Q^* is not exactly the logarithm of a density function means that, unlike pseudo-likelihood, the extended quasi-likelihood estimate $\hat{\theta}_{QL}$ of θ is not always consistent. For Poisson data with small means the inconsistency should be noticeable. Such examples have small expected cell counts. Morton (1987a) also concludes that the extended quasi-likelihood estimate can be unreliable in these cases, and he says that methods based on squared residuals 'will often be preferable'.

3.3.6 General mean–variance functions using replication

In many assay problems the experimenters replicate the response at each value of the predictor. The predictor levels are $(x_1, \ldots, x_N)$, and at each value x_i we observe m_i replicated responses $(y_{i,j})$. Write the means and variances as

$$E(y_{i,j}) = \mu_i = \mu_i(\beta)$$
$$\text{Standard deviation of } y_{i,j} = \sigma_i$$

With no other information, it is tempting to estimate β using weighted least squares where the estimated weight is the inverse of the sample variance at each level of the predictor, i.e.

$$\hat{w}_i = 1.0/s_i^2$$

$$s_i^2 = (m_i - 1)^{-1} \sum_{j=1}^{m_i} (y_{i,j} - \bar{y}_i)^2$$

where $\bar{y}_i$ is the sample mean at predictor value x_i. Often the number of replicates is rather small, with two or three typical. In this case, it is well known that the resulting weighted least-squares estimator can be a disaster. Jacquez et al. (1968) and Jacquez and Norusis (1973) point this out in Monte Carlo studies. An asymptotic theory is given by Carroll and Cline (1988), who note that this weighted least-squares estimate will be inconsistent for $m = 2$ replicates. Simple asymptotic theory shows that for normally distributed errors with a constant number of replicates, the replicated variance estimator does not follow the asymptotic theory of section 2.2, but is instead more variable than one would expect. For example, even $m = 6$ replicates is three times more variable than what one expects, while $m = 10$ replicates leads to a variance inflation of 40%. See Cochran (1937) and Yates and Cochran (1938) for illuminating discussions. Fuller and Rao (1978) show that this same phenomenon persists even when the sample means are replaced by ordinary least-squares predicted values, although in this case the variance inflation typically is not as big a problem, especially when the number of replicates exceeds six. The trouble with the replicated variance estimator is that the weights are poorly estimated. The sample variances are multiples of chi-squared random variables with few degrees of freedom, and hence they can vary wildly; it is instructive to inspect tables of chi-squared percentage points divided by degrees of freedom to see the difference between using two and 30 observations to estimate variance. Using only the replicated sample variances to estimate weights is equivalent in spirit to estimating the regression function only by the plot of predictors against group means without fitting any model. It is not too surprising that such a 'connect-the-dots' strategy will not work very well for usual sample sizes.

One way to use the sample variances smoothly is to replace absolute residuals by sample standard deviations and apply one of the previous fitting methods. Such substitution is treated theoretically in

Davidian and Carroll (1987). When the variances do not depend on the mean or if σ is small, the effect of this substitution on pseudo-likelihood is to lose some efficiency. Assume that the number of replicates at each design point is m. The asymptotic efficiency of using sample variances in pseudo-likelihood rather than squared residuals is

$$(m - 1)(2 + \kappa)[2 + (m - 1)(2 + \kappa)]^{-1}$$

where κ is the kurtosis of the errors and is zero at the normal distribution. Thus, the efficiency is 50% for duplicates at the normal distribution, while for quadruplicates it is 75%. For weighted absolute errors the figures are 50% and 82% for duplicates and quadruplicates, respectively.

For other transformations, sample standard deviations need not be less efficient at the normal distribution. This is particularly true for logarithms (see Davidian and Carroll, 1987).

One advantage of using the sample standard deviations concerns model robustness. The sample standard deviations give us information about the variance function even if the model for the means is incorrectly specified, whereas in this case using absolute residuals would be misleading.

There are two other methods in use which assume replication and a model for the variances, although not for the means. These methods are standard in assay work. In many of these specific applications, the difference between sample means and fitted means from a regression function is miniscule, rarely larger than 0.5%. It makes sense in these situations to assume that σ is very small, so that the effect of estimating the regression function can be ignored. As in (2.1), we assume that the standard deviations are

$$\sigma g(\mu_i, \theta)$$

but for the moment we will not assume that we know the regression function $f(x, \beta)$. This is not as unreasonable as it may look on first sight. In many assays we observe not only both the predictors and the responses, but also an additional set of data where the response but not the predictor is observed. Since this second set may make up a substantial proportion of the data, it makes sense to use it to get a better estimate of the variance parameter θ. Of course, one could use the predictors for those observations where they are measured, which would be more efficient than never using the predictor.

Methods based on regression The most widely used regression-based method is due to Rodbard and his colleagues; see Rodbard and Frazier (1975) and Rodbard (1978) for examples. This is a close analog to the logarithm method discussed earlier in this section. It is hard to overstate the impact this work has had on statistical practice in assays, because it has enabled investigators to avoid dependence on unweighted methods. Consider the power-of-the-mean model (2.5), and note that taking logarithms results in a linear model with slope θ

$$\log(\sigma_i) = \log(\sigma) + \theta \log(\mu_i)$$

This leads to the suggestion that one might regress the logarithm of the sample standard deviation s_i on the logarithm of the observed sample mean $\bar{y}_i$. The resulting estimate is not a consistent estimate of θ for two reasons. The first, often rather minor in assays, is that the sample means are not the true means, so that regressing on the logarithm of the sample means is an errors-in-variables problem; see Fuller (1987) for a thorough discussion of bias in errors-in-variables models. The second problem can arise if the number of replicates is not constant across groups. As Raab (1981a) has pointed out, the expected value of the logarithm of the sample standard deviation is linear in the logarithm of the true means but with an intercept depending on the number of replicates. For example, for normally distributed data

$m_i = 2$	implies	$E \log(s_i) = 0.635 + \log(\sigma) + \theta \log(\mu_i)$
$m_i = 3$	implies	$E \log(s_i) = 0.288 + \log(\sigma) + \theta \log(\mu_i)$
$m_i = 4$	implies	$E \log(s_i) = 0.182 + \log(\sigma) + \theta \log(\mu_i)$

For normally distributed data one should subtract the biasing qualities on the right side of these equations.

It is perhaps not too surprising that since the Rodbard estimate of θ does not use the regression function, it is inefficient with respect to pseudo-likelihood. Of course, this inefficiency is relative and the Rodbard estimate is far superior to using unweighted methods. In order to compute the efficiency of Rodbard's estimate, we make the simplifying assumption that the number of replicates is constant across groups, that σ is small, and that the sample size is large. These calculations have been performed by Davidian and Carroll (1987) and are given in Table 3.6.

Table 3.6 suggests a simple way to improve upon Rodbard's

Table 3.6 *The asymptotic efficiency of Rodbard's estimate of θ with respect to pseudo-likelihood in the power-of-the-mean model*

Number of replicates	Efficiency (%) of Rodbard's method
2	20.2
3	40.6
4	53.4
5	62.0
6	68.0
7	72.4

original idea when there are data points in which only the response is available. Suppose that there are M_1 such data points, and that there are M_2 pairs that are completely observed. Assume that all points have m replicates, and let c_{eff} denote the efficiency as given in Table 3.6. If $\hat{\theta}_R$ is Rodbard's estimate and $\hat{\theta}_{\text{PL}}$ is pseudo-likelihood, then a weighted estimate of θ is

$$\frac{M_2\hat{\theta}_{\text{PL}} + c_{\text{eff}}M_1\hat{\theta}_R}{M_2 + c_{\text{eff}}M_1}$$

Since Rodbard's method is closely related to the logarithm method, we conjecture that Rodbard's method is more robust to heavy tailed distributions than is pseudo-likelihood.

Maximum-likelihood-based methods   We again consider the power-of-the-mean model (2.5). Raab (1981a) estimates θ, σ, and $(\mu_1, \ldots, \mu_N)$ simultaneously by maximizing a modified normal likelihood. For $\theta = 0$ the usual maximum-likelihood estimate of σ is biased, but of course it can be made unbiased by dividing the corrected sum of squares by the degrees of freedom rather than the sample size. Alternatively, one can modify the likelihood to achieve the same end; the modified loglikelihood used by Raab is

$$-\sum_{i=1}^{N}(m_i - 1)\log(\sigma\mu_i^\theta) - \tfrac{1}{2}\sigma^{-2}\sum_{i=1}^{N}\sum_{j=1}^{m_i}[(y_{i,j} - \mu_i)/\mu_i^\theta]^2$$

This function is to be maximized jointly in θ, σ, and $(\mu_1, \ldots, \mu_N)$. Sadler and Smith (1985) suggest replacing μ_i in this expression by the sample means, a practice we would support in their context. These estimates are closely related to a nonlinear maximum-likelihood estimate in an errors-in-variables problem (Fuller, 1987), and the number of parameters $(N + 2)$ increases with the sample size. Fuller's book reveals that this is a difficult problem theoretically. However, for large sample sizes and small values of σ, Davidian and Carroll (1987) have shown that the efficiency of the Raab and Sadler–Smith estimates $\hat{\theta}_{EIV}$ with respect to pseudo-likelihood is $(m - 1)/m$ in the case of equal replication with m replicates. For purposes of comparison, in Table 3.7 these values are listed along with the efficiencies of Rodbard's estimate. By dividing the second column of this table by the third, we obtain the efficiencies of Rodbard's estimator with respect to the Raab and Sadler–Smith methods, being 40.4%, 60.9%, and 71.2% for $m = 2, 3$, and 4, respectively. These theoretical efficiencies are similar to those reported in a Monte Carlo study by Raab (Raab, 1981a).

If one has M_2 pairs of responses and concentrations which are completely observed and M_1 pairs in which only the response is observed, then we suggest using the weighted estimator of θ given by

$$\frac{M_2 \hat{\theta}_{PL} + (m - 1)M_1 \hat{\theta}_{EIV}/m}{M_2 + (m - 1)M_1/m}$$

Table 3.7   *The asymptotic efficiences of the Rodbard, Raab, and Sadler–Smith estimates of θ with respect to pseudo-likelihood in the power-of-the-mean model*

Number of replicates	Efficiency (%) of Rodbard method	Efficiency (%) of Raab and Sadler–Smith methods
2	20.2	50.0
3	40.6	66.7
4	53.4	75.0
5	62.0	80.0
6	68.0	83.3
7	72.4	85.7

3.3.7 Comments

One must be careful in applying any of the methods for very small sample sizes. We have seen attempts to estimate simultaneously a three-parameter regression function and a power-of-the-mean model for the standard deviations based on 10 observations, but even with bias-corrected pseudo-likelihood and normally distributed data we wonder how well any of the parameters can be estimated. In such a situation, a bootstrap experiment could be very sobering.

While methods such as weighted regression on absolute residuals or regression on the logarithm of absolute residuals are qualitatively more robust than pseudo-likelihood, such robustness is relative. In fact, none of these methods can be described as robust in the modern sense. A discussion of robustness and diagnostics is given in Chapter 6.

If there are replicates at each design point, one might wish to replace absolute or squared residuals by sample standard deviations and variances, respectively. As we have seen, this alternative typically is a little less efficient for normally distributed data but is sometimes easier to compute. As discussed above and mentioned by Raab (1981a), problems arise if the number of replicates at each design point is not constant. A discussion of the theory is given in Davidian and Carroll (1987).

There is one theoretical curiosity that we have finessed. The normal-theory maximum-likelihood estimate of the regression parameter β was discussed in the previous chapter and we do not recommend its routine use when the variance is a function of the mean response. The normal-theory maximum-likelihood estimate of θ is a pseudo-likelihood estimate based on the maximum-likelihood estimate of β. Davidian (1986) has shown in this case that the normal-theory maximum-likelihood estimate of θ is asymptotically more efficient than pseudo-likelihood based on a generalized least-squares estimate of β even at distributions other than the normal, although for most of the problems we encounter, the increase in efficiency would be minimal.

It is difficult and probably foolish to recommend a single method of variance-function estimation. Pseudo-likelihood is based on either squared residuals or sample variances, involves only standard calculations, and has the advantages of making few assumptions and extending readily to more complicated variance structures such as occur in components-of-variance models. All of the methods can be greatly influenced by an individual observation or a small group of observations, so that they should not be used without some sort of influence diagnostic. See Chapter 6 for more details.

3.4 Inference

In this section we will discuss some methods for making inferences about the parameters β and θ in the heteroscedastic regression model. In subsections 3.4.1–3.4.4 we discuss inference about θ when the effect of estimating β can be reasonably ignored. In subsection 3.4.5, we discuss Wald-type inference for β and θ based on the theory of M-estimation discussed in section 7.1.

The standard and easiest way to make inferences about the regression parameter β is to fix θ at its estimated value and apply any of the methods of section 2.5. These ignore the extra variability caused by estimation of θ. To capture this extra variability one can use bootstrap inference, θ estimated simultaneously with β and σ. See Efron and Tibshirani (1986) and Efron (1987) for a discussion of bootstrap inference (see also section 2.5).

For nearly normally distributed data, estimating θ generally causes the usual asymptotic theory to be more optimistic than we would prefer, and Wald inference and the usual confidence intervals ought to be adjusted, at least in principle. One possible method for correcting the Wald test statistic is through the use of variance expansions. Letting α denote an individual component of the regression parameter β with estimated standard error formed by the usual asymptotic theory, the t-statistic will be written in the form

$$(\hat{\alpha}_G - \alpha)/(\text{Estimated standard error})$$

In principle at least, one can compute an expansion for the variance of this test statistic, the expansion being of the form

$$\text{Variance}(t\text{-statistic}) \simeq 1 + c_\alpha/N$$

One could then estimate the constant c_α and divide the t-statistic by the square root of $(1 + \hat{c}_\alpha/N)$. Unfortunately, obtaining a formula for c_α is not easy even under the best of circumstances. As suggested in the previous chapter, this correction is automatically accomplished by using a bootstrap when the bootstrap idea is applicable. Also, $\hat{\alpha}$ will not be symmetrically distributed in general, and the bootstrap makes an asymmetric correction.

Inference about the parameter θ has been discussed by many authors; see Judge et al. (1985) who list 23 references on this problem. See also Cook and Weisberg (1983), Carroll and Ruppert (1981b), Koenker and Bassett (1981), Bickel (1978), and Harrison and McCabe (1979). A full discussion might be longer than our treatment of estimation. Instead we will only discuss a few methods that can be applied rather generally. To save on notation we assume throughout that the errors are independent and identically distributed.

For basic background, the results of Davidian and Carroll (1987) include the following. If the regression parameter β were known, then the asymptotic theory for estimating θ would have a particularly simple form. Define $v(i, \beta, \theta)$ as in (3.9), v_θ its derivative with respect to θ, and $\xi(\beta, \theta)$ the sample covariance matrix of v_θ, i.e.

$$\xi(\beta, \theta) = \left(N^{-1} \sum_{i=1}^{N} \left[v_\theta(i, \beta, \theta) - \bar{v}_\theta(\beta, \theta) \right] \right.$$

$$\left. \times \left[v_\theta(i, \beta, \theta) - \bar{v}_\theta(\beta, \theta) \right]^{\mathrm{T}} \right) \tag{3.26}$$

Then, as shown by Davidian and Carroll (1987), pseudo-likelihood, restricted maximum-likelihood, and weighted squared residual estimates of θ are asymptotically normally distributed with mean θ and covariance

$$\frac{2 + \kappa}{4N} \xi(\beta, \theta)^{-1} \tag{3.27}$$

where κ is the average kurtosis of the errors (2.21), which need not be identically distributed. Regression based on logarithms of absolute residuals is asymptotically normally distributed with mean θ and covariance

$$\frac{\text{Variance}[\ln(\varepsilon^2)]}{4N} \xi(\beta, \theta)^{-1} \tag{3.28}$$

Regression based on weighted absolute residuals is asymptotically normally distributed with mean θ and covariance

$$\frac{\delta}{(1 - \delta)N} \xi(\beta, \theta)^{-1} \tag{3.29}$$

where

$$\delta = \text{Variance}(|\varepsilon|).$$

3.4.1 Inference for variance parameters based on asymptotic standard errors

The asymptotic distribution theory of estimating θ is complicated by the possible dependence of the asymptotic distribution on the estimate of β. Suppose that θ has q components. An approximation is available if either the variance function does not depend on the mean and thus can be written as $g(z_i, \theta)$, or if σ is small. In these cases, as shown by Davidian and Carroll (1987), pseudo-likelihood and weighted squared residual estimates of θ are asymptotically normally distributed with mean θ and approximate covariance matrix (3.27). Estimated standard errors for the components of θ can be formed by direct estimation of the quantities in (3.27), but this is tedious. A more direct approach is to remember that pseudo-likelihood estimation of θ is equivalent to minimizing the weighted sum of squares based on the problem (3.18). Thus approximate standard errors can be obtained by reading them off from nonlinear least-squares programs with

$$\text{'Responses'} = [y_i - f(x_i, \hat{\beta}_*)]^2$$
$$\text{'Weights'} = g^{-4}(\mu_i(\hat{\beta}_*), z_i, \hat{\theta}_{\text{PL}})$$
$$\text{'Regression Function'} = \sigma^2 g^2(\mu_i(\hat{\beta}_*), z_i, \theta)$$

We have not made an empirical study of the performance of Wald confidence intervals and tests for the parameter θ. As a first approximation, the estimated standard errors are easy to compute, which may be looked upon as an advantage.

Inference based on weighted absolute residual estimation or the logarithms of absolute residuals (Harvey's method) can be performed in much the same way. These can both be computed as regression problems. As an approximation for σ small, one can use the standard errors from the computer output. There is a problem with this approach if the errors (2.21) are asymmetric, in which case there is an effect due to estimating β (see Davidian and Carroll, 1987).

3.4.2 Inference based on pseudo- and restricted maximum likelihood

The log-pseudo-likelihood $L_{\text{PL}}(\beta, \theta)$ of the data as a function of β and θ is given by (3.8). It is tempting to perform inference pretending that this is actually the logarithm of the likelihood. If θ is a $(q \times 1)$ vector and we wish to test whether $\theta = \theta_0$, and if $\hat{\beta}_G$ and $\hat{\beta}_G(\theta_0)$ are generalized least-squares estimates calculated at $\hat{\theta}_{PL}$ and θ_0 respectively, then two reasonable test statistics are

$$T_1 = 2[L_{PL}(\hat{\beta}_G, \hat{\theta}_{PL}) - L_{PL}(\hat{\beta}_G(\theta_0), \theta_0)]$$

$$T_2 = 2[L_{PL}(\hat{\beta}_G, \hat{\theta}_{PL}) - L_{PL}(\hat{\beta}_G, \theta_0)] \qquad (3.30)$$

If the variance function does not depend on the mean and can be written as $g(z_i, \theta)$ or if σ is small, then under the null hypothesis, these two test statistics satisfy

$$\frac{2}{2 + \kappa} T_j \approx \chi_q^2 \qquad \text{for } j = 1, 2$$

where κ is the kurtosis of the errors (2.21) and $\approx$ signifies convergence in distribution. Letting $\hat{\kappa}$ be the estimated kurtosis defined by

$$\hat{\kappa} = (N - \rho)^{-1} \sum_{i=1}^{N} \{e_i^2(\hat{\beta}_G, \hat{\theta}_{PL}) - 1\}^2 - 2 \qquad (3.31)$$

this suggests multiplying the two test statistics by $2/(2 + \hat{\kappa})$ and comparing the result with percentage points of a chi-squared distribution having q degrees of freedom. Since κ is usually poorly estimated except for large sample sizes, we tend to be wary of making this correction. The effect of ignoring this correction when the kurtosis is larger than zero is to inflate the level of the test and to form optimistic confidence intervals, i.e., intervals that are too small. Neither of these are necessarily bad at the data analysis stage, but should be kept in mind before rejecting a proposed value of θ.

Let $\hat{\sigma}^2(\beta, \theta)$ be defined by (3.7). Let $S_G(\beta, \theta)$ be given by (2.4). From (3.10), the restricted loglikelihood is given by

$$L_{REML}(\beta, \theta) = L_{PL}(\beta, \theta) + \tfrac{1}{2} \log_e \{ \text{Det} [\hat{\sigma}^2(\beta, \theta) S_G^{-1}(\beta, \theta)] \}$$

$$(3.32)$$

Since the correction to the pseudo-likelihood is negligible for large sample sizes, a test that takes into account the loss of degrees of freedom for estimating β is (3.30) using L_{REML} rather than L_{PL}.

3.4.3 Score tests for the variance parameter

Score tests have been suggested by Cook and Weisberg (1983) and Breusch and Pagan (1979). Strictly speaking their work only applies to testing the presence of heteroscedasticity and not to testing for a particular type of heteroscedasticity. Suppose we wish to test the hypothesis that the value of θ is θ_0. Assume for the moment that the observations are normally distributed and β and σ are known, and write $L(\beta, \theta, \sigma)$ as the logarithm of the likelihood for the data. The form of the score statistic is given in Chapter 7. This form is rather complicated in general but simplifies in important special cases. It is possible to find an explicit expression for the score statistic which is applicable in three cases:

(1) The variances do not depend on the mean.

(2) The variance depends on the mean but the derivative with respect to β of the logarithm of g evaluated at θ_0 is zero.

(3) The variance depends on the mean, no special value of θ_0 is considered but as an approximation σ is considered 'small'. More precisely, the asymptotic theory is large N, small σ.

The second case arises when one is testing for the presence of any heterogeneity. For example, in the power-of-the-mean model (2.5), $\theta_0 = 0$ means constant variance and it is at this value of θ_0 for which (2) is satisfied. Define $v(i, \beta, \theta)$ as in (3.9) and define the standardized residuals by

$$e_i^2(\beta, \theta) = \frac{[y_i - f(x_i, \beta)]^2}{\sigma^2(\beta, \theta)g^2(\mu_i(\beta), z_i, \theta)} \tag{3.33}$$

where $\sigma^2(\beta, \theta)$ is defined in (3.7). Let $\hat{\beta}_G(\theta_0)$ be a generalized least-squares estimate of β assuming θ_0 is the correct value of θ and using full iteration, i.e., quasi-likelihood. Then under any of the three special cases discussed previously, the score statistic can be written as $\Omega(\hat{\beta}_G(\theta_0), \theta_0)$, where

$$\Omega(\beta, \theta) = Q_1^T Q_2^{-1} Q_1 \tag{3.34}$$

$$Q_1(\beta, \theta) = N^{-1/2} \sum_{i=1}^{N} [e_i^2(\beta, \theta) - 1]d(i, \beta, \theta) \tag{3.35}$$

$$Q_2(\beta, \theta) = 2N^{-1} \sum_{i=1}^{N} d(i, \beta, \theta)d(i, \beta, \theta)^T \tag{3.36}$$

and

$$d(i, \beta, \theta) = v_\theta(i, \beta, \theta) - \bar{v}_\theta(\beta, \theta)$$

Here $v(i, \beta, \theta)$ is defined by (3.9), v_θ is the derivative with respect to θ, and $\bar{v}_\theta$ is the average value of v_θ. Assuming the hypothesis is true, $\Omega(\hat{\beta}_G(\theta_0), \theta_0)$ has an asymptotic chi-squared distribution with q degrees of freedom, where q is the number of elements making up θ. One would reject the hypothesis based on the percentage points of this distribution.

The test statistic (3.34) is essentially equivalent to that of Cook and Weisberg (1983). To define a test with the correct asymptotic level for all distributions, the leading constant 2 in (3.36) should be replaced by $(2 + \kappa)$, where κ is the kurtosis of the errors (2.21). For nonnormal distributions this can be estimated by $(2 + \hat{\kappa})$, where $\hat{\kappa}$ is given by (3.31). If one is convinced that the data are very nearly normally distributed, then this substitution should not be made because estimating the kurtosis is difficult. See McCullagh and Pregibon (1987) for a further discussion.

The score statistic has one particularly nice feature. Suppose that we decide to model the standard deviation as proportional to $h(z^T\theta)$ for some predictors z and unknowns θ. It turns out that the score statistic for the hypothesis of homogeneous variance $\theta = 0$ does not depend on the function h.

When θ is a scalar rather than a vector, the statistic (3.34) has a simple interpretation. In this case, if the leading constant 2 in (3.36) is replaced by $(2 + \hat{\kappa})$ and $(N - p)$ in (3.31) is replaced by N, the test statistic is N times the squared Pearson correlation between squared standardized residuals (3.33) and the quantities $d(i, \beta, \theta)$. If we want to test for the presence of heteroscedasticity against power-of-the-mean alternatives, these latter quantities are the logarithm of the predicted values. If the test is for heteroscedasticity with an alternative of the form

$$\exp[\theta f(x, \beta)]$$

then we would use simple predicted values.

Thus, for scalar θ simple approximate tests can be based on the Pearson correlation between squared standardized residuals at θ_0 and predicted values or their logarithms. By analogy, for a quick approximate test absolute residuals could be used in place of squared residuals, and Spearman correlations might be used in place of Pearson correlation.

It is important to note that the score test can be greatly affected by a single point, especially by an observation that has small predicted variability under the hypothesis. We have seen examples that the score test statistic had value 4.0 or more for the full data, but 1.0 or less when a single point out of nearly 100 is deleted. Instead of using an extremely nonrobust diagnostic for heteroscedasticity in data analysis, we prefer to use a simple robust alternative such as the Spearman correlation.

3.4.4 Inference based on the extended quasi-likelihood

The quasi-likelihood ratio test for whether $\theta = \theta_0$ is complicated to study theoretically because of the general inconsistency of the estimate of θ. Davidian and Carroll (1988) give an analytic description of this phenomenon in the small σ case. From (3.24) the extended quasi-likelihood test statistic is

$$-2 \sum_{i=1}^{N} [l_Q^*(\mu_i(\hat{\beta}_G(\theta_0)), y_i, \theta_0, \hat{\sigma}(\theta_0))]$$

$$+2 \sum_{i=1}^{N} [l_Q^*(\mu_i(\hat{\beta}_G(\hat{\theta}_{QL})), y_i, \hat{\theta}_{QL}, \hat{\sigma}(\hat{\theta}_{QL}))]$$

We have found that while the other estimates of θ discussed in section 3.3 are often very similar, the behavior of the quasi-likelihood test statistic can be markedly different from them. We will illustrate this point in Example 3.4 in section 3.5. It appears that the quasi-likelihood test is sensitive to the shape of the distribution indexed by θ_0, as well as being influenced by the variance function itself. Davidian and Carroll (1988) also point out that test levels may be affected by the inconsistency of the estimate of θ, even for small σ.

3.4.5 Wald-type inference based on M-estimation

The methods of estimating the parameters $\xi = (\beta, \sigma, \theta)$ which we have discussed are M-estimators (section 7.1), solving an equation of the form

$$0 = \sum_{i=1}^{N} \Psi(y_i, x_i, \xi) \tag{3.37}$$

(see equation (7.2)). For example, when fully iterating pseudo-likelihood the function Ψ is given by combining equations (2.7) and (3.10). Since solutions to (3.37) are M-estimates, their joint asymptotic distribution is given by equation (7.9), i.e.

$$N^{1/2}(\hat{\xi} - \xi) \to \text{Normal}(0, B^{-1}A(B^{-1})^{\mathrm{T}})$$

Letting Ψ_ξ be the derivative of Ψ with respect to ξ, from (7.11) and (7.12) the matrices A and B can be estimated by

$$\hat{A} = N^{-1} \sum_{i=1}^{N} \Psi(y_i, x_i, \hat{\xi}) \Psi(y_i, x, \hat{\xi})^{\mathrm{T}}$$

$$\hat{B} = N^{-1} \sum_{i=1}^{N} \Psi_\xi(y_i, x_i, \hat{\xi})$$

Wald-type inference (section 7.2) can be based on the approximation that $\hat{\xi}$ is asymptotically normally distributed with mean θ and covariance matrix given by

$$N^{-1}\hat{B}^{-1}\hat{A}(\hat{B}^{-1})^{\mathrm{T}}$$

3.5   Examples

In this section we illustrate numerically some of the methods discussed in the previous three sections. Standard errors for estimates of θ were computed as in subsection 3.4.1.

Example 3.1   Oxidation of benzene

This is Example 2.1 (section 2.8), for which graphical analyses suggest a constant-coefficient-of-variation model. Pritchard *et al.* (1977) fit the power-of-the-mean model (2.5) to the full data using normal-theory maximum likelihood and obtained $\hat{\theta} \simeq 1.15$.

We first regressed the absolute residuals from an unweighted least-squares fit on the 'regression function'

$$af(x, \hat{\beta})^\theta \tag{3.38}$$

where $\hat{\beta}$ is the unweighted least-squares estimate. This gave an estimated value $\hat{\theta} = 1.12$. However, when we then weighted these residuals as suggested in section 3.3, the updated estimate was $\theta = 0.41$. This extremely unstable behavior might be an example of the result that how well one estimates β has an effect on how well one can estimate θ. More concretely, there is one observation that has a rather large absolute residual and a rather large predicted value. The effect of weighting this point in the absolute residual analysis could account for some of the instability. We note that having fixed $\hat{\beta}$, the parameters (a, θ) in (3.38) are estimated by generalized least squares, and in keeping with our discussions it makes sense to update the estimates iteratively. Convergence was very slow, with a final estimate of $\hat{\theta} = 0.82$. For example, even on the tenth cycle on the process, the changes in the estimate of θ were of absolute size 0.20. The culprit here appears to be the unweighted least-squares residuals, combined with the poor initial guess $\theta = 0$ to begin the iterative process. This is the kind of behavior observed by Matloff *et al.* (1984).

We next used the residuals from fully iterated (quasi-likelihood) generalized least squares with $\theta = 0.80$. The step sizes in the iteration for estimating θ were much smaller, with rapid convergence to $\theta = 1.05$. We repeated these steps, obtaining a final estimate $\hat{\theta} = 1.13$, with an estimated standard error 0.16 computed as in subsection 3.4.1. This is an interesting example because the computational difficulties we experienced using unweighted least-squares residuals disappeared when we combined the residuals from a sensible weighted fit with reasonable starting values for θ.

The pseudo-likelihood estimate was computed on the grid of values between 0 and 1.5 in steps of 0.10. We started with unweighted least-squares residuals, obtained a first estimate of θ, then used the residuals from this value of θ to obtain an updated estimate, etc. This process converged to $\hat{\theta}_{PL} = 1.10$. The confidence interval using the likelihood test T_1 in section 3.4 was from 0.90 to 1.40, essentially the same as obtained using weighted absolute residuals. Extended quasi-likelihood computed on the same grid gave a similar analysis.

When we regressed the logarithm of the absolute residuals from an unweighted least-squares fit against the logarithm of the predicted values having deleted the two smallest of the former, we obtained $\hat{\theta} = 0.68$ with a standard error of 0.21. We then began an iterative process, using the residuals from fully iterated or quasi-likelihood generalized least squares based on $\hat{\theta} = 0.68$, obtaining a new estimate of θ, which lead to new residuals, etc. Rounded, the process converged to $\hat{\theta} = 0.85$, with a standard error of 0.20. There is a slight discrepancy here with the previous methods, but a plot of the logarithm of the absolute residuals versus the logarithm of the predicted values as well as an influence analysis as in Chapter 6 suggests that two observations are having a large impact on the estimate. These are the data points with responses 9.44 and 9.15. For illustrative purposes we deleted these two observations, obtaining the new estimate (rounded) $\hat{\theta} = 1.10$.

Example 3.2　Rat data

This data set is discussed by Weisberg (1985, pp. 121–4) and by Cook (1986). The data consist of 19 cases and the regression model is linear based on three predictors: ξ_1 = body weight, ξ_2 = liver weight, and ξ_3 = relative dose. The response was the percentage of the dose in the liver. The experiment was set up to show that, within the range of the design, there was no relationship between the response and the predictors. It is our understanding from a discussion with Professor Cook that the experimenter had rather strong views about the lack of any relationship. As discussed in Weisberg (1985), there is a dramatic change in the unweighted least-squares linear regression analysis depending on whether case 3 is retained in the analysis. Since Weisberg raises some questions as to the validity of this case, we have deleted it. The unweighted least-squares analysis indicates no predictive ability from a linear model: for example, the R-square is 0.02.

We next examined absolute studentized and Anscombe residual plots against the predicted values as well as each of the predictors. There certainly was no monotone relationship here, with all the

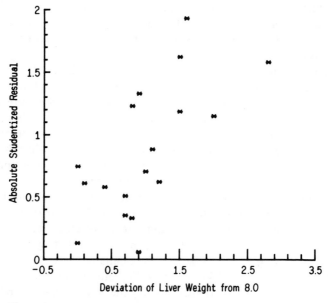

Figure 3.1　*Rat data: unweighted least-squares fit.*

Spearman rank correlations being small and insignificant. Cook (1986) suggests that there may be some heterogeneity of variance as a function of liver weight. An examination of the absolute studentized residuals suggests a possible quadratic variance function in $\xi_2 =$ liver weight (see Figure 3.2). By our eyeball estimate, Figure 3.2 looks roughly symmetric about the deviations on liver weight from 8.0; we will call this quantity the liver weight score. When we plot the absolute studentized residuals against the liver weight score in Figure 3.1, there is a strong and definite pattern, which also appears with the Anscombe residuals. The Spearman rank correlation in this figure is $\rho = 0.66$ with approximate significance level 0.003.

The finding of possible heterogeneity of variance in liver weight with essentially constant mean is unexpected. Owing to the nature of the experiment, we would question whether this relationship is reproducible. However, in the following numerical illustration we will tentatively assume that the standard deviations are quadratic in liver weight. Thus our model is

$$f(x, \beta) = \beta_0 + \beta_1\xi_1 + \beta_2\xi_2 + \beta_3\xi_3 \qquad (3.39)$$

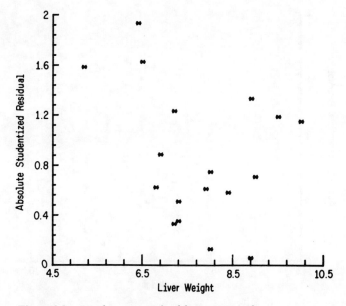

Figure 3.2   *Rat data: unweighted least-squares fit.*

$$\text{Standard deviation}(y) = \theta_0 + \theta_1(\xi_2 - 8.0) + \theta_2(\xi_2 - 8.0)^2 \quad (3.40)$$

An inspection of Figure 3.1 suggests that $\theta_1 = 0$, at least approximately.

Since we are modeling the variance as a function of predictors, the maximum-likelihood estimate is also a pseudo-likelihood estimate. When listing the estimates, for scaling purposes we normalized the estimates by multiplying them by 125. The estimates, along with the standard errors computed from the regression program, are

$$\hat{\theta}_0 = 0.41 \quad (0.08) \qquad \hat{\theta}_1 = -0.01 \quad (0.08) \qquad \hat{\theta}_2 = 0.23 \quad (0.08)$$

We also fit the same standard-deviation function by using weighted absolute residuals. The parameter estimates and their standard errors were

$$\hat{\theta}_0 = 0.28 \quad (0.07) \qquad \hat{\theta}_1 = -0.03 \quad (0.08) \qquad \hat{\theta}_2 = 0.24 \quad (0.08)$$

The difference between these two estimates is that the maximum-likelihood estimate gives less weight to those observations with liver weight near 8.0. Using the maximum-likelihood estimate, the absolute studentized residual plot after weighting is given in Figure 3.3.

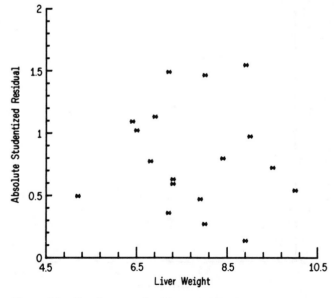

Figure 3.3   *Rat data: weighted least-squares fit.*

There is no discernible pattern, and now the Spearman rank correlation with the liver weight score is essentially zero.

An alternative model also suggested by Figure 3.1 is one in which the standard deviations are linear in the liver weight score.

Example 3.3   Car insurance data

The car insurance data were introduced in section 2.7. The observed responses were the average insurance claims as a function of policy-holder's age group, car group, and vehicle age group, which we will refer to as PA, CG, and VA, respectively. Write y_{ijk} to be the observed average claim when (PA, CG, VA) = (i, j, k), and let n_{ijk} represent the number of observations upon which y_{ijk} is based. As a numerical illustration only, we fit a model in which the means were given by

$$Ey_{ijk} = (\mu + \alpha_i + \beta_j + \gamma_k)^{-1}$$

Here α_i, β_j, and γ_k refer to the usual analysis-of-variance effects. We took the standard deviation to follow the power-of-the-mean model (2.5) divided by the square root of the number of observations making up that observation. As described in section 3.4, the computed standard errors can be taken only as rough guides.

We first regressed the logarithm of the absolute residuals from a least-squares fit on the logarithm of the predicted values, after having trimmed the smallest five absolute residuals (see sections 2.7 and 3.4). The estimate of the power parameter was $\hat{\theta} = 1.63$, with a computed standard error of 0.27. When we did this using the quasi-likelihood residuals from fits with $\theta = 1.00$ and $\theta = 1.50$, essentially the same answer was obtained.

We next performed a weighted nonlinear regression on the absolute residuals. We first used unweighted least-squares residuals

$$r_{ijk}^* = (n_{ijk})^{1/2}(y_{ijk} - \hat{y}_{ijk})$$

and estimated θ by regressing $|r_{ijk}^*|$ on the function (2.5), without weighting. The estimate of θ obtained in this way was $\hat{\theta} = 1.11$. We then weighted these absolute residuals as suggested in section 3.3 and then iterated the process to convergence, obtaining a weighted estimate $\hat{\theta} = 1.30$ with standard error 0.24. We then computed the residuals r_{ijk}^* from a quasi-likelihood fit with $\theta = 1.30$ and computed an updated iterated estimate $\hat{\theta} = 1.46$, with a standard error of 0.25.

We next computed a pseudo-likelihood estimate of θ, restricting our calculations to values of $\theta = 0, 0.1, \ldots, 1.5$. Starting from unweighted least-squares residuals, the estimate was $\hat{\theta} = 1.10$. We used this value of θ to generate an updated estimate of β, then updated θ, etc. The process converged to $\hat{\theta}_{PL} = 1.30$. The restricted-likelihood estimate based on degrees-of-freedom corrections converged to the same value. By any of the testing devices suggested in section 3.4, values of θ less than 0.80 are rejected at the 5% level, with $\theta = 1.00$ having approximately the same chi-squared value as $\theta = 1.50$. This is somewhat in contrast to the other regression methods and the Spearman correlations, which tend to favor $\theta = 1.50$.

The extended quasi-likelihood estimate of θ on this grid is 1.20. There is a substantial difference with the previous methods, in that the extended quasi-likelihood chi-squared test for $\theta = 1.00$ has value 2.13 while that for $\theta = 1.50$ has value 6.75. The interesting point here is that the extended quasi-likelihood test seems to operate entirely differently from the others.

In summary, the estimates of θ lie in the range 1.20 to 1.50. Regression methods and Spearman correlations based on absolute residuals tend to favor $\theta = 1.50$ over $\theta = 1.00$, while pseudo-likelihood and restricted likelihood show no difference between them, and quasi-likelihood favors $\theta = 1.00$. All the methods agree that choosing $\theta \leqslant 0.80$ would be a mistake.

A preliminary influence analysis (see Chapter 6) indicates that at least three combinations of (PA, CG, VA) might be exerting an unsual effect on the estimates of θ, these combinations being (2, 2, 1), (3, 4, 4), and (3, 2, 4). We repeated the analysis having deleted these three observations. The Spearman correlation between absolute studentized residuals and predicted values was 0.23 for $\theta = 1.00$ and 0.06 for $\theta = 1.50$. Regression based on the logarithm of absolute residuals yields an estimate of $\hat{\theta} = 1.59$ with a standard error of 0.27. Regression based on weighted absolute residuals yields essentially the same result: $\hat{\theta} = 1.64$ with standard error 0.22. The pseudo-likelihood estimate is $\hat{\theta}_{PL} = 1.50$, while the restricted-likelihood estimate is essentially the same; all testing methods based on these estimates reject $\theta = 1.00$. All these results differ from extended quasi-likelihood, which estimates θ as 1.20 and rejects $\theta = 1.50$.

The substantial differences between extended quasi-likelihood and all the other methods are interesting. What may be happening here is that extended quasi-likelihood is picking up a distributional difference between its distribution for $\theta = 1.0$ and $\theta = 1.5$. In this particular example, such differences are probably not too important. Instead of prediction limits, estimates of the mean and standard errors for the mean seem of prime importance here and do not depend too strongly on whether one chooses $\theta = 1.00$ or $\theta = 1.50$.

Example 3.4   Tensile strength data

Box and Meyer (1985b) present an interesting analysis of dispersion effects in a fractional-factorial experiment. The experiment concerns the tensile strength of welds in an off-line welding experiment performed by the National Railway Corporation of Japan (Taguchi and Wu, 1980). As described by Box and Meyer (1985b), in this screening experiment it is desired to uncover which factors have large effects on the outcome. Box and Meyer (1985a) concluded that the means could be adequately explained by two factors, B and C. The raw data are plotted in Figure 3.4; we have arranged the plot so that all the data are exhibited. There are clear effects on the mean due to factors B and C, and there is little hint of a significant interaction.

As noted by Box and Meyer, Figure 3.4 also shows a difference in variability due to the two levels of factor C. As an illustration, we first

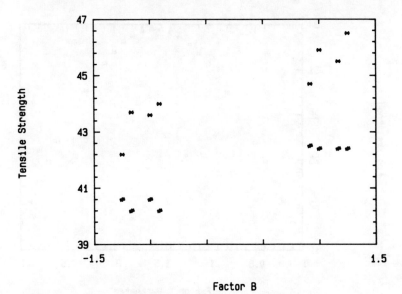

Figure 3.4   *Tensile strength data: symbols are levels of factor C.*

fit the same model as did Box and Meyer, namely a loglinear model

$$\text{Variance}(y_i) = \exp(\theta_0 + \theta_C C_i) \qquad (3.41)$$

where $C_i = +1$ or -1. The method of fitting was maximum
likelihood, which is a pseudo-likelihood estimate since the variance is
not being modeled as a function of the mean response. The maximum-
likelihood estimate of θ_C was approximately 1.55. This is quite a large
effect, as it suggests that the standard deviations vary over the levels of
C by a factor of almost 5. A plot of the logarithm of the likelihood
function in Figure 3.5 suggests a 95% confidence interval of approximately 1.00 to 2.00.

To test for an effect on variances due to factor B, we fit the expanded
loglinear model

$$\text{Variance}(y_i) = \exp(\theta_0 + \theta_B B_i + \theta_C C_i) \qquad (3.42)$$

The maximum-likelihood estimate of θ_C now became 2.00, while the
maximum-likelihood estimate for θ_B was -0.80. These estimates
would suggest that changing the levels of factor C changes the
standard deviations by a proportional factor of 7, while doing the
same for B changes the standard deviations by a proportional factor

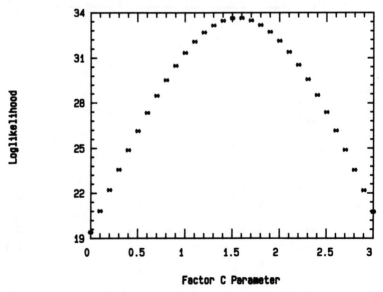

Figure 3.5   *Tensile strength data: loglikelihood with no B effect.*

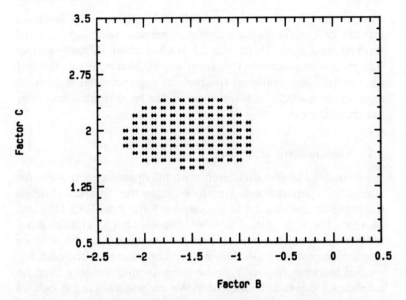

Figure 3.6   *Tensile strength data: 95% confidence set; factors B and C interact for variance.*

of 2, which seems to conform with Figure 3.4. A 95% likelihood ratio confidence ellipse for θ_B and θ_C is traced in Figure 3.6, and note that the hypothesis of no effect due to B is marginally significant at the 5% level. The effect of factor C on the variances is still quite major. We were surprised by the statistical significance of factor B, which appears in retrospect to be due to the very small variability of the response when both B and C are at level -1. This suggests that we should not have an additive loglinear model for the variances as in (3.42) but rather the effect of factor B should only appear when factor C is at level -1. We do not pursue this issue, especially since it is not clear to us whether the effect, if real, is at all important.

We had some interesting computational experiences with this example. One direct method of computing the maximum-likelihood estimate is to start with the least-squares estimate $\hat{\beta}_L$ from an unweighted fit to an additive model, and then compute the maximum-likelihood estimate of θ assuming that $\beta = \hat{\beta}_L$. We then update $\hat{\beta}$ by generalized least squares, obtaining a new estimate of θ, etc. This process converged very quickly. Alternatively, instead of direct computation of the maximum-likelihood estimate of θ given an estimate of β, one could use weighted regression methods on squared residuals as suggested in section 3.3. We had terrific difficulty getting this process to converge. This surprised us, because it was the first time we had any significant trouble with regression-based comput-ation of the estimate of θ which could not be attributed to a poor parameterization.

3.6 Nonparametric smoothing

We have dealt almost exclusively with fully parametric models. An alternative approach would be to estimate the variances through nonparametric regression techniques (see Silverman, 1985; Hall and Marron, 1987). In this section we discuss what is known about estimating the regression parameter β when the form of the variance function has not been specified. We first concentrate on the case that generalized least squares is to be used, so that we only need an estimate of the variance function. We do not discuss the case of maximum-likelihood rather than generalized least-squares estim-ation, which is in the emerging literature of semiparametric adapt-ation. References in this latter area include Bickel (1982), Begun et al. (1983), and Carroll et al. (1986b). The literature on semiparametric maximum-likelihood estimation is undergoing a vigorous expansion and will be better covered in a few years when we have a perspective on the results.

As we have seen in section 2.7, nonparametric estimation of the variance function via smoothing techniques can be a useful data analytic device. We now look at the effect of smoothing on generalized least-squares estimates of β.

Recall the basic asymptotic result for a fully parametric specific-ation of the variance function as in (2.1). At least in the limit, generalized least-squares estimates are just as efficient as weighted least-squares estimates with known weights. Now suppose that a parametric specification of variance is not known, and that instead we are only sure that *for an unknown function* g_0 either

$$\text{Standard deviation of } y_i = g_0(x_i) \tag{3.43}$$

or

$$\text{Standard deviation of } y_i = g_0(\mu_i(\beta)) \tag{3.44}$$

The question is whether it is possible under either (3.43) or (3.44) to construct an estimator of β which does as well as if the weights were known. If we can do this, then we will say that we can adapt for heteroscedasticity.

One possible construct has been investigated in an important paper by Fuller and Rao (1978); see also Carroll and Cline (1988). Suppose that at each design point x_i there are m_i replicate observations. In practice, the number of replicates is usually small. A method often used in practice is this. First fit unweighted least squares to the data and then compute the absolute deviations $d_{ij}(\hat{\beta}_{LS})$, where

$$d_{ij}(\beta) = |y_{ij} - f(x_i, \beta)| \qquad (3.45)$$

As an estimate of the variance at x_i take

$$\hat{\sigma}_i^2 = m_i^{-1} \sum_{j=1}^{m_i} d_{ij}^2(\hat{\beta}_{LS}) \qquad (3.46)$$

Estimated weights $\hat{w}_i = 1.0/\hat{\sigma}_i^2$ are formed and generalized least squares applied. This is a 'connect-the-dots' variance-function estimator. For the most part these weights are unreliable because they are based on a small number of degrees of freedom. Technically, the 'connect-the-dots' method yields an inconsistent estimate of the variance function, so it should come as no surprise that the resulting estimate of β does not adapt. This was proved for the normal distribution by Fuller and Rao (1978). The increase in asymptotic variance for β due to using this inconsistent variance function can be substantial. The estimated variances (3.46) are not smooth enough, and this appears to be the cause of the poor behavior of the resulting generalized least-squares estimate of β; see also Rao and Subrahmanian (1971).

Carroll (1982a) noted that there is more information about the variance function than simply the estimates (3.46). The chain of reasoning begins with the observation that, in regression on means, for values of x 'near' one another the true means are typically assumed to be similar. Since we have argued previously that variance-function estimation is really a form of regression, it makes sense to take as a working hypothesis that for values of x 'near' one another the true variances are typically similar. This argument suggests estimating the variance at x by a form of weighted averaging of the squared residuals, those design points closest to x getting the most weight. Such averaging is the basic idea of most nonparametric regression methods. The idea was that a variance-function estimate obtained by

smoothing, being a consistent estimate, would be just good enough to allow us to estimate β well.
One method of estimating the variance function is to note that

$$\text{Variance}(y_i) = E(y_i^2|x_i) - [E(y_i|x_i)]^2$$

This suggests two applications of nonparametric regression by estimating the conditional expectations of both y_i^2 and y_i given x_i. If one is confident that the mean function is $f(x_i, \beta)$, then it is more efficient to implement a single application of nonparametric regression using the relationship

$$\text{Variance}(y_i) = E(d_i^2(\beta)|x_i) \tag{3.47}$$

The predictor x_i is a vector of p elements, the jth of which will be denoted $x_i^{(j)}$. Assuming that β is known and that the variances follow the model (3.43), consider estimating $g_0(x)$ by a weighted average of the terms $d_i^2(\beta)$, with largest weight given to those observations closest to x, i.e.,

$$\hat{g}_0^2(x, \beta) = \sum_{i=1}^{N} c_{i,N}(x)d_i^2(\beta) \tag{3.48}$$

where

$$\sum_{i=1}^{N} c_{i,N}(x) = 1$$

In order to obtain a consistent estimate of the variance function, as the sample size gets large, $c_{i,N}(x)$ should approach a Dirac function evaluated at x. One way to make this operational is through the use of kernel functions. Let $b(N) \to 0$ control the amount of local averaging, and let K be a unimodal p-dimensional density function with mode at the origin. Define

$$c_{i,N}(x) = a_{i,N}(x) \bigg/ \sum_{j=1}^{N} a_{j,N}(x) \tag{3.49}$$

where

$$a_{i,N}(x) = K\left(\frac{x_i^{(1)} - x^{(1)}}{b(N)}, \ldots, \frac{x_i^{(p)} - x^{(p)}}{b(N)}\right)$$

Note how $b(n)$ controls the amount of averaging. Larger values of $b(N)$ imply more smoothness but also some bias, so that as $b(N) \to \infty$

for fixed N we get

$$\hat{g}_0^2(x_i, \beta) \to N^{-1} \sum_{j=1}^{N} d_j^2(\beta)$$

which is a constant and clearly unacceptable. Smaller values of $b(N)$ imply less bias but also less smoothness and more variance, with $b(N) \to 0$ for N fixed being equivalent to 'connecting the dots'. Clearly how we choose the bandwidth matters greatly; see Haerdle et al. (1987), Hall and Marron (1987), and Marron (1986, 1987). Here is our suggestion. Let $\hat{\beta}$ be an estimator of β. One possibility is unweighted least squares, but we would prefer not to use this because it is so sensitive to leverage. Because of heteroscedasticity, unweighted least squares tends to find 'outliers' at the high leverage points and is thus a particularly gruesome method on which to base one's estimate of weights. Form the kernel variance function $\hat{g}_0^2(x, \hat{\beta})$, and then form a weight function

$$\hat{w}(x, \hat{\beta}) = [\hat{g}_0^2(x, \hat{\beta}) + \eta_N]^{-1} \qquad (3.50)$$

Here η_N is a deterministic positive sequence of numbers converging to zero more slowly than the bandwidth; its main value is to eliminate the possibility of infinite weights. We will denote by $\hat{\beta}_w$ the generalized least-squares estimate of β based on (3.50).

The following result appears to hold in some generality. However, the research is still in a rather primitive state, focusing on such simple problems as linear regression.

Result 3.1 Using the weighting scheme (3.50), the estimate $\hat{\beta}_w$ adapts for heteroscedasticity, i.e., has the same asymptotic distribution as if the weights were known.

This very encouraging result was first discovered by Carroll (1982a), although proved only for the special case of linear regression with a $(p = 1)$-dimensional predictor. The result has been generalized by Muller and Stadtmuller (1986) and Robinson (1986). The proofs tend to be long and technical. For a hopeful Monte Carlo study, see Matloff et al. (1984).

Based on our experience with least-squares residuals, we recommend that the variance function be re-estimated by using the deviations $d_i(\hat{\beta}_w)$. Generalized least squares would then be applied based on the re-estimated variance function. Further iteration might be considered, although we have no experience with doing so.

Model (3.44) postulates that the variances are strictly a function of the mean response. There is an unfortunate and unwise tendency to dismiss this model because it is formally a special case of (3.43). Occam's razor dictates that simpler models can be more easily understood and estimated. If there are 10 predictors and the variance is usefully modeled as a function of the mean, it makes no sense to do 10-dimensional nonparametric regression. The kernel regression estimator is changed only by replacing x in the argument for (3.49) by $f(x, \beta)$. Carroll (1982a) has shown that an analog of Result 3.1 holds, and thus we can adapt for heteroscedasticity if the variances are known only to depend on the mean.

In this subsection, we have indicated methods for smoothing squared residuals. The resulting estimated variance functions are efficient enough that in large samples for estimating β one can treat the weights as if they were known. In sections 2.7 and 3.3, we also advocated the smoothing of absolute residuals and their logarithms. Robust smoothers might also be used. We know of no theory for these alternative methods.

Since by smoothing squared residuals we can estimate β asymptotically as well as if the variances were known, there is a tendency to believe that finding a parametric model for the variance function is not worth the effort. We take the position that parametric and nonparametric methods are both worthwhile. Parametric models are easier to understand and more efficient to estimate. Nonparametric methods have an important role to play in uncovering structure, especially for problems with moderately large sample sizes. The major importance of the results in this section is that they indicate the basic gain in efficiency for estimating β when one exploits the simple, basic idea that variance-function estimation is a form of regression.

CHAPTER 4

The transform-both-sides
methodology

4.1 Introduction

In nonlinear regression, a model $f(x, \beta)$ for the expectation of the response y is often available from a theoretical understanding of the system giving rise to the data, through preliminary data analysis, or through experience with similar data. A model for the random variation of y about $f(x, \beta)$ is less commonly available. For example, to study pulmonary defense mechanisms, mice can be placed in a chamber containing an aerosol of bacteria labeled with a radiotracer (Ruppert *et al.*, 1975). At specified times, the mice are sacrificed and the lungs are removed and homogenized. The total quantity of bacteria inhaled and the quantity still viable at sacrifice are measured by placing the homogenate in a scintillation counter and by culturing a sample of the lung homogenate. Let x_t, x_0, and t be the measured number of viable bacteria after sacrifice, the measured number inhaled, and the time between exposure and sacrifice. Let $y = (x_t/x_0)$ $\times 100\%$ be the percentage of bacteria viable at time t. Assuming a constant bacterial death rate in the lung, the bacteria will decrease exponentially. This leads to the model

$$y = \exp(\alpha + \beta t) \qquad (4.1)$$

The bacterial death rate is $-\beta$, and α should be approximately $\log(100)$. There are numerous reasons why this model will not hold exactly. In particular there will be measurement errors in y, x_0, and t. Further β varies in time and, more importantly, across animals. In this example, none of the sources of variation are easy to model. A model combining all of them seems impossible given the type and amount of data available. It appears necessary to model empirically the fluctuations in y about (4.1).

It is common practice simply to assume additive, normal errors in (4.1), i.e., that

$$y = f(x, \beta) + \varepsilon \qquad (4.2)$$

where ε is normally distributed with mean 0 and variance σ^2. Unfortunately, the residuals often are heteroscedastic and/or skewed and may contain outliers. Heteroscedasticity alone can be handled by the weighting techniques discussed in the previous chapters, but weighting by reciprocal variances is alone not suitable for highly nonnormal ε. In Chapter 6 we discuss an entirely different method of weighting, where the weights are a decreasing function of the 'influence' of an observation. These so-called robust estimators can accommodate certain types of nonnormality, in particular the presence of outliers. However, most robust estimators assume a symmetric distribution.

There are two well established ways of modeling skewed data. One approach lets the density of ε be in a parametric class of skewed densities, say gamma densities. This leads to generalized linear models (McCullagh and Nelder, 1983). Our approach assumes that y can be transformed to a symmetric distribution. Both generalized linear models and transformation models have been used successfully in practice. Both can, in fact must, be generalized to cover the variety of data found in practice. For example, generalized linear models can be extended to generalized *nonlinear* models. Variance functions, discussed in the previous chapter, can be used with either generalized linear or transformation models.

Let $h(y, \lambda)$ be a family of transformations of y indexed by the parameter λ. In many examples λ is scalar, but this is not necessary. It is generally required that $h(y, \lambda)$ be monotone in y; otherwise, a model for $h(y, \lambda)$ cannot generate a model for y by inverting the transformation.

The most commonly used family is the modified power transformations of Box and Cox (1964)

$$h(y, \lambda) = y^{(\lambda)} = \begin{cases} (y^\lambda - 1)/\lambda & \text{if } \lambda \neq 0 \\ \log y & \text{if } \lambda = 0 \end{cases} \qquad (4.3)$$

Another possibility is a power transformation with shift

$$h(y, \lambda, \mu) = (y - \mu)^{(\lambda)} \qquad (4.4)$$

The use of (4.3) assumes that y is positive. If y has a minimum possible value, then $(y - \mu)$ can be made positive by the proper choice of μ. The shifted power transformation is discussed in detail by Atkinson (1985).

After choosing the transformation family, (4.2) is replaced by

$$h(y, \lambda) = h(f(x, \beta), \lambda) + \varepsilon \qquad (4.5)$$

Model (4.5) will be called 'transform both sides' and abbreviated by TBS. The model states that, after applying the same transformation to y and $f(x, \beta)$, the residuals are normally distributed with a constant variance (Carroll and Ruppert, 1984a).

The motivation for (4.5) is as follows. When there are no sources of variation, that is when $\varepsilon \equiv 0$, (4.2) and (4.5) are equivalent. We transform the response y to induce symmetric errors with a constant variance. Then we apply the same transformation to $f(x, \beta)$ to preserve the relationship between y and x. Equation (4.5) allows random variation to enter the model in a variety of forms. Moreover, if $h(y, \lambda)$ is linear in y for some λ, then (4.5) includes (4.2) as a special case; one has the option not to transform when the data so indicate. For example, when using the modified power transformation family, we can test the null hypothesis that no transformation is necessary by testing $H_0: \lambda = 1$.

If $h(y, \lambda)$ is a sufficiently rich family, for example the modified power transformations, then (4.5) can be fit successfully to data sets exhibiting severe heteroscedasticity and/or nonnormality. The parameter λ can be estimated along with β and σ, so that the data themselves can determine how we model the effect of the disturbance ε on the response y.

It should be stressed that transformation model (4.5) is very different from the model studied by Box and Cox (1964). In their model the response y is transformed but not the regression function $f(x, \beta)$. The Box–Cox transformation model is most appropriate when $f(x, \beta)$ is selected for its simple form, say straight-line regression, but $f(x, \beta)$ does not fit the untransformed y. The transformation h is chosen so that $f(x, \beta)$ fits $h(y, \lambda)$. It is hoped that the same h induces normality and a constant variance.

Our transformation model assumes that the regression equation $y = f(x, \beta)$ already fits the data adequately, but the residual variation is nonnormal or heteroscedastic. This assumption is reasonable whenever a model $f(x, \beta)$ has already been chosen empirically, as in Snee (1986), or else is available from theoretical considerations. We transform both y and $f(x, \beta)$ to preserve the regression equation. When the residual variation exhibits skewness or heteroscedasticity then several reasons exist for using the transform-both-sides method rather than simply fitting y to $f(x, \beta)$ by unweighted least squares. First, transformation to normality and homoscedasticity allows the parameter vector β to be estimated efficiently. Without a transformation, the ordinary least-squares estimator of β will have an unnecessarily large variance. Moreover, the TBS model provides a model of the entire conditional distribution of y given x. Therefore, confidence intervals for the mean or quantiles of y are available as well as prediction intervals for y and calibration intervals for x given y. Even when $\hat{\beta}$ estimates β well, confidence, prediction, and calibration intervals based upon ordinary least squares can be grossly in error if the residual variation is nonnormal or has a nonconstant variance; this point has already been discussed in Chapter 2.

In section 4.2 we discuss the modified power transformation family further, in particular how it can accommodate heteroscedasticity and skewness in the original response y. Section 4.3 discusses estimation of β, λ, and σ. Estimation of the conditional mean and the conditional quantiles of y given x is the topic of section 4.4. Several examples are treated in detail in section 4.5.

Other aspects of transformations are discussed in later chapters. Although transformations can remove heteroscedasticity and nonnormality, for a given data set there is no guarantee that one transformation can do both. It may be necessary simultaneously to transform to remove skewness and to weight to stabilize the variance. In Chapter 5 weighting and transformations are compared, and methods are introduced for deciding when to transform, when to weight, or when to do both. Outliers can have drastic effects on estimated transformation parameters. For example, one outlier can induce tremendous skewness or apparent heteroscedasticity in an otherwise normal and homoscedastic set of data. Conversely, right-skewness can be masked by a single outlier in the left tail, and heteroscedasticity may be hidden if an outlier occurs at an observation that would otherwise have a small variance. In Chapter 6 robust estimation and diagnostics for influential and outlying points are introduced for transformation models.

The idea of transforming both sides of a regression model has been around for a long time. Its traditional use has been to linearize otherwise nonlinear models; there are many examples where a simple transformation of $f(x, \beta)$, say the inverse or logarithm, is linear in the parameters. Modern nonlinear software makes such transformation unnecessary. Moreover, linearizing transformations may induce asymmetry or heteroscedasticity, making ordinary least squares very inefficient. Box and Hill (1974) give an example where the effect of linearization is severe induced heteroscedasticity and a physically impossible estimate of a parameter. They suggest retaining the transformation and correcting the induced heteroscedasticity by weighting. Carroll and Ruppert (1984a) show that an equally satisfactory fit results from using a transformation to homoscedasticity.

Transforming both sides to achieve normality and a stable variance has no doubt occurred to data analysts from time to time. Snee and Irr (1981) used this idea in a mutagenesis study. The present authors became aware of the technique during a project modeling the Atlantic menhaden population. A biomathematician, R.B. Deriso, suggested log-transforming both sides of a catch–effort model to eliminate the potential effects of severe right-outliers in catch data, in effect inducing near-symmetry. Later in the same project when modeling spawner–recruit data, we used TBS with the parameter estimated by maximum likelihood.

Carroll and Ruppert (1984a) appear to be the first to advocate transform both sides as a general statistical technique.

Another early application of transform both sides in a special case appears in Leech (1975). He was concerned with testing whether the errors entered the model additively, corresponding to $\lambda = 1$ in the modified power transformation family, or multiplicatively, corresponding to $\lambda = 0$. Leech suggested that one could give meaning to other values of λ, but he did not pursue this suggestion.

Another useful model is the linear regression of the transformed response $y^{(\lambda)}$ on the transformed explanatory variable $x^{(\lambda)}$, i.e.

$$y^{(\lambda)} = \alpha_0 + \alpha_1 x^{(\lambda)} + \varepsilon \qquad (4.6)$$

This model is often appropriate when x is a lagged value of y and is related to the common practice of transforming a time series before fitting an ARIMA (autoregressive integrated moving average) model; see Hopwood et al. (1984) and Box and Jenkins (1976, p. 303) for examples. Egy and Lahiri (1979) use (4.6) with heteroscedastic errors; their model is closely related to the model in Chapter 5. Model (4.6) can also be used when y and x are different measurements of the same quantity. For example, Leurgans (1980) takes log-transformations when comparing a test and reference method of measuring glucose in the blood. Model (4.6) is similar to the transform-both-sides model

$$y^{(\lambda)} = (\beta_0 + \beta_1 x)^{(\lambda)} + \varepsilon \qquad (4.7)$$

used empirically by Snee (1986). When α_0 and β_0 are 0, then (4.6) and (4.7) are equivalent. An advantage of (4.7) is that β_1 is a physically meaningful parameter even when λ is unknown, since $\beta_0 + \beta_1 x$ is the conditional median of y given x. By (4.6), the conditional median of y is

$$h^{-1}[\alpha_0 + \alpha_1 x^{(\lambda)}, \lambda]$$

which depends on the unknown λ. Here $h^{-1}[y, \lambda]$ is the inverse of $h[y, \lambda]$ as a function of y with λ fixed, i.e., $h^{-1}[h(y, \lambda), \lambda] = y$.

There are sampling situations where the transform-both-sides method is equivalent to the 'transform-the-response' approach. For example, in the k-sample problem or one-way layout, one has a sample of size n_i from population i, $i = 1, \ldots, k$. The transform-both-sides model is

$$h(y_{ij}, \lambda) = h(\mu_i, \lambda) + \varepsilon_{ij} \qquad (4.8)$$

where y_{ij} is the jth observation from the ith population and the ε are independent normal variates with a constant variance. The transform-the-response model is

$$h(y_{ij}, \lambda) = \theta_i + \varepsilon_{ij} \qquad (4.9)$$

The only difference between (4.8) and (4.9) is in the meaning of μ_i and θ_i. In (4.8), μ_i is the median of y_{ij}, while in (4.9) θ_i is the median and the mean of $h(y_{ij}, \lambda)$. Notice that $\theta_i = h(\mu_i, \lambda)$, and the maximum-likelihood estimates satisfy the same relationship by their invariance under reparameterizations. Thus we can translate the result from the transform-both-sides model to the transform-the-response model, or vice versa, without re-estimating. Also, both models lead to the same estimates of the distribution of y given x.

Since the null hypothesis $H_0: \mu_1 = \ldots = \mu_k$ is true if and only if the null hypothesis $H_0: \theta_1 = \ldots = \theta_k$ is true, when testing for a group effect the models are equivalent.

For estimation purposes the two models behave differently. The $\hat{\theta}$ in model (4.9) will have large variances and are highly correlated with $\hat{\lambda}$; see Bickel and Doksum (1981), Box and Cox (1982), and section 4.3.

When using model (4.9), Hinkley and Runger (1984) suggest conditioning on $\hat{\lambda}$. The conditional variance of $\hat{\theta}_i$ given $\hat{\lambda}$ is small, but the 'parameter' being estimated is $E(\hat{\theta}_i | \hat{\lambda})$ which is random! In model (4.8) the $\hat{\mu}$ have low sampling variability and are not highly correlated with $\hat{\lambda}$. This is related to the fact that μ_i has physical meaning even when λ is unknown. The conditional approach of Hinkley and Runger does not seem necessary if (4.8) is used.

Another introduction to the TBS model is given by Snee (1986). His excellent article contains a number of interesting applications.

There are several papers about the Box–Cox 'transform-the-response' method that discuss issues of relevance to transform both sides. Carroll (1982c) gives an example where outliers have a tremendous effect on estimated transformation parameters. His paper demonstrates the need for diagnostics and robust procedures. Earlier papers on robust estimation in the Box–Cox model include Andrews (1971), Carroll (1980), and Bickel and Doksum (1981). Carroll (1982b) discusses the effects on requiring $\hat{\lambda}$ to be in a restricted set, e.g., $\{-1, -1/2, 0, 1/2, 1\}$, which is common practice. Carroll (1983) and Doksum and Wong (1983) discuss testing hypotheses about the regression parameters in the context of transformation. Carroll and Ruppert (1981a, 1984b) discuss prediction of the untransformed y, in particular estimation of the conditional median of y given x.

4.2 Heteroscedasticity, skewness, and transformations

The intelligent use of transformations requires understanding of their effects upon nonnormality and heterogeneity of variance. This section shows how transformations remove heteroscedasticity when the variance is a function of the mean and how the convexity, or concavity, of a transformation determines its effects upon skewness.

Suppose that the random variable y_i has a mean μ_i and a variance σ_i^2 that are functionally related so that $\sigma_i = \sigma g(\mu_i)$ for some function g and constant σ. As shown in Bartlett (1947), if y_i is transformed to $h(y_i)$ then the variance of $h(y_i)$ can be approximated by a simple Taylor series expansion

$$\begin{aligned} \text{Var}(h(y_i)) &\simeq E[h(y_i) - h(\mu_i)]^2 \\ &\simeq (\dot{h}(\mu_i))^2 E[y_i - \mu_i]^2 \\ &= (\dot{h}(\mu_i))^2 [\sigma g(\mu_i)]^2 \end{aligned} \qquad (4.10)$$

where $\dot{h}$ is the derivative of h. Therefore the variance of $h(y_i)$ is approximately constant if $\dot{h}(y)$ is proportional to $(g(y))^{-1}$. Typically g will be positive and then h will be monotonically increasing. Monotonicity of h makes interpretation of the transformation easier since certain qualitative relationships between y and the independent variables, e.g., that y is an increasing function of a particular one of the independent variables, will be preserved by monotonic transformations. Moreover, if an objective is a prediction model for y, then it is essential that h be an invertible transformation.

The accuracy of the Taylor series approximation depends upon the distance between y_i and $f(x_i, \beta)$ and the degree of nonlinearity of h. In general, the approximation is accurate when y_i has a small variance or the second derivative of h is small compared to the first derivative.

There are many important examples where g is a power function, and then h can also be a power or modified power transformation. For example, $g(\mu) = \mu^{1/2}$ for Poisson-distributed data, and then $h(\mu)$ should be a linear function of $\mu^{1/2}$ so one can take $\lambda = 1/2$ in the Box–Cox family. If $g(\mu)$ is equal to μ, then the coefficient of variation is constant and $h(\mu)$ should be proportional to μ^{-1}; log-transformation stabilizes the variance. When the shape parameter is constant, lognormal- and gamma-distributed data are special cases. In general, when $g(\mu) = \mu^{(1-\lambda)}$ then $\dot{h}$ should be proportional to $\mu^{(\lambda-1)}$, and either the modified power transformation $h(y, \lambda) = y^{(\lambda)}$ or, when $\lambda \neq 0$, the ordinary power transformation is appropriate.

Van Zwet (1964) has introduced a definition of relative (right-) skewness. He defines the distribution function G to be more (right-) skewed than the distribution function F if $R(t) = G^{-1}(F(t))$ is a convex function. To see the reasoning behind this definition, note that if X is F-distributed then $Y = R(X)$ is G-distributed. Since the convex function R has an increasing first derivative, the transformation R from X to Y tends to 'push together' values in the left tail and 'pull apart' values in the right tail relative to its effects on central values. Compared to X, Y will have a longer right tail and a shorter left tail. Therefore, van Zwet's definition is a quite natural translation of an intuitive notion of skewness into a precise mathematical notion. Van Zwet has shown that the ordinary standardized third-moment coefficient of skewness γ respects his definition of right-skewness, in that $\gamma(G) \geqslant \gamma(F)$ if $G^{-1}F$ is convex.

According to van Zwet's definition, if we transform Y to $\varphi(Y)$ then we increase right-skewness if φ is convex, while we decrease right-skewness if φ is concave. Therefore, $\varphi(Y)$ will have a greater or smaller skewness coefficient than Y according to whether φ is convex or concave.

In the Box–Cox family $y^{(\lambda)}$ is convex for $\lambda > 1$ and concave for $\lambda < 1$. In our experience right-skewed data are more common than left-skewed, and this explains why λ is typically less than 1 in practice. However, left-skewed data sets exist, and to induce approximate symmetry they should be transformed by λ greater than 1. In many biologic and economic data sets, the response is positive, right-skewed, and heteroscedastic with the variance an increasing function of the mean. In such a case, a transform $y^{(\lambda)}$ with $\lambda < 1$ will reduce both skewness and heteroscedasticity, though there need not be a single value of λ transforming to both symmetry and constant variance.

In the modified power transformation family, $y^{(\lambda)}$ more severely affects skewness and heteroscedasticity as λ moves away from 1. If λ is fixed in the modified power transformation family with shift μ, then $h(y, \lambda, \mu) = (y - \mu)^{(\lambda)}$ becomes more severe as μ tends toward the minimum of $y_1, \ldots, y_N$.

Bickel and Doksum (1981) mention that, for $\lambda > 0$, $y^{(\lambda)}$ can be extended to negative y by the transformation

$$y \to [\,|y|^{\lambda}\mathrm{sgn}(y) - 1]/\lambda$$

This transformation changes from convex to concave as y changes sign, so its effect on skewed data will be difficult to predict. Such concave–convex function can be applied to a symmetric distribution to change the kurtosis (van Zwet 1964, Oja 1981, Ruppert 1987). If $\lambda < 1$, then the effect is to decrease peakedness and tail weight so that the kurtosis is reduced.

When the variance of y_i is small relative to the variation among the means $\mu_1, \ldots, \mu_N$, then the transformation $h(y_i)$ is essentially equivalent to the transformation $[h(\mu_i) + \dot{h}(\mu_i)(y_i - \mu_i)]$. Being linear in y_i, this transformation has no effect on distributional shape, but because the slope varies with i, it has the effect on the mean–variance relationship shown by equation (4.10). Thus when 'σ is small', the transform-both-sides model is indistinguishable from the heteroscedastic model

$$y_i = \mu_i + \sigma\dot{h}(\mu_i)\varepsilon_i$$

discussed in Chapters 2 and 3.

The variance need not be too small for the two models to be nearly equivalent. As mentioned in the previous section, Box and Hill (1974) use a heteroscedastic model, with the variance proportional to a power of the mean, to analyze chemical kinetics data. Although the variability in that example is not small, Carroll and Ruppert (1984a) found that applying a power transformation 'to both sides' leads to the same conclusions and nearly identical estimates of the regression parameters.

When the means $\mu_1, \ldots, \mu_N$ are nearly constant then transformation will have little or no effect on heteroscedasticity. Transformations can still be of importance to induce a symmetric, nearly normal distribution; see Example 4.2 in section 4.5.

4.3 Estimation and inference

Suppose $h(y, \lambda)$ is a family of transformations indexed by an m-dimensional vector λ and for some λ the transform-both-sides model

$$h(y, \lambda) = h(f(x, \beta), \lambda) + \varepsilon \qquad (4.11)$$

holds, where the ε_i are independent, identically distributed according to the distribution function F. The first topic of this section is joint estimation of β, λ, and σ by maximum likelihood, assuming that $F = N(0, \sigma^2)$, the normal distribution with mean 0 and variance σ^2. The case of general $h(y, \lambda)$ will be briefly discussed. We discuss in detail the case where $h(y, \lambda)$ is the modified power transformation $y^{(\lambda)}$. Then the likelihood function can be re-expressed so that standard nonlinear least squares software will compute the maximum-likelihood estimates. The second topic is calculation of standard errors and confidence regions for λ and β. Finally we discuss estimation methods other than maximum likelihood. These methods attempt to transform to homoscedasticity, symmetry, or both and do not depend on a normality assumption.

4.3.1 Maximum-likelihood estimation

Let the Jacobian of the transformation $y_i \to h(y_i, \lambda)$ be $J_i(\lambda)$, i.e.

$$J_i(\lambda) = dh(y, \lambda)/dy|_{y_i}$$

so that $J_i(\lambda) = y^{\lambda - 1}$ if $h(y, \lambda)$ is a modified power transformation. Then the conditional density of y_i given x_i is

$$f(y_i|x_i, \beta, \lambda, \sigma) = (2\pi\sigma^2)^{-1/2} \exp\{-[h(y_i, \lambda) $$
$$- h(f(x_i, \beta), \lambda)]^2/(2\sigma^2)\}J_i(\lambda)$$

and the loglikelihood for $y_1, \ldots, y_N$ given $x_1, \ldots, x_N$ is, up to an additive constant,

$$L(\beta, \lambda, \sigma) = -N\log(\sigma) - \sum_{i=1}^{N}[h(y_i, \lambda) - h(f(x_i, \beta), \lambda)]^2/(2\sigma^2)$$

$$+ \sum_{i=1}^{N}\log[J_i(\lambda)] \qquad (4.12)$$

For fixed β and λ, $L(\beta, \lambda, \sigma)$ is maximized in σ by

$$\hat{\sigma}^2(\beta, \lambda) = \sum_{i=1}^{N}[h(y_i, \lambda) - h(f(x_i, \beta), \lambda)]^2/N \qquad (4.13)$$

and the MLE of β and λ maximizes

$$L_{max}(\beta, \lambda) = L(\beta, \lambda, \hat{\sigma}(\beta, \lambda))$$

$$= -N\log[\hat{\sigma}(\beta, \lambda)] - N/2 + \sum_{i=1}^{N}\log[J_i(\lambda)]$$

$$= -(N/2)\log\{\hat{\sigma}(\beta, \lambda)^2/[\dot{J}(\lambda)]^2\} - N/2 \qquad (4.14)$$

where

$$\dot{J}(\lambda) = \left(\prod_{i=1}^{N}J_i(\lambda)\right)^{1/N}$$

is the geometric mean of $J_1(\lambda), \ldots, J_N(\lambda)$. The subscript 'max' means that the function is maximized over the omitted variables, here σ.

$L_{max}(\beta, \lambda)$, or even $L(\beta, \lambda, \sigma)$, can be maximized using an iterative optimization technique such as a Newton or quasi-Newton algorithm, or the related Fisher method of scoring. The latter was used in the examples in this and the next two chapters. Such methods are convenient when available and have certain advantages when estimating standard errors (see below). Unfortunately, many statistical packages do not yet have general maximum-likelihood routines.

Nonlinear least-squares software is now readily available. When using the modified power transformations, $L_{max}(\beta, \lambda)$ can be put in a form suitable for least-squares estimation. As Box and Cox (1964) note, the Jacobian is $J_i(\lambda) = y_i^{\lambda-1}$ and by easy algebra (4.14) becomes

$$L_{\max}(\beta, \lambda) = -(N/2)\log[\hat{\sigma}^2(\beta,\lambda)/\dot{y}^{2(\lambda-1)}] - N/2 \qquad (4.15)$$

where $\dot{y}$ is the geometric mean of $y_1, \ldots, y_N$. Therefore $\hat{\beta}$ and $\hat{\lambda}$ minimize

$$\sum_{i=1}^{N} \{[y_i^{(\lambda)} - f^{(\lambda)}(x_i, \beta)]/\dot{y}^\lambda\}^2 \qquad (4.16)$$

Expression (4.16) can be minimized by standard nonlinear least-squares software. Most nonlinear regression routines do not allow the response to depend upon unknown parameters, so unless λ is fixed we *cannot* use $y_i^{(\lambda)}/\dot{y}^\lambda$ as the response and $f^{(\lambda)}/\dot{y}^\lambda$ as the model. Instead we reformulate the model so that the response does not depend on parameters. To do this we create the dummy variable D_i which is identically equal to zero and use D_i as the 'pseudo-response'. Then we incorporate the true response into a 'pseudo-model'. Define the 'pseudo-regression function'

$$e(x_i, \beta, \lambda) = [y_i^{(\lambda)} - f^{(\lambda)}(x_i, \beta)]/\dot{y}^\lambda$$

with 'independent' variables x and y and 'regression' parameters β and λ. The 'pseudo-model' fits D_i to $e(x_i, \beta, \lambda)$

$$D_i = e(x_i, \beta, \lambda) \qquad (4.17)$$

The residual from the pseudo-model is $e(x_i, \beta, \lambda)$, which is the same residual as the transformation model; this is the reason for setting the 'response' D_i identically zero. The least-squares estimates from this pseudo-model minimize (4.16) and therefore are the MLEs even though (4.17) is not a *bona fide* regression model. The MLE of σ^2 can be recovered from (4.13), that is $\hat{\sigma}^2 = \hat{\sigma}^2(\hat{\beta}, \hat{\lambda})$. To correct for degrees of freedom, we use

$$s^2 = N/(N-p-1)\hat{\sigma}^2 \qquad (4.18)$$

Another method of computing the maximum-likelihood estimates with standard software exists but is not quite as convenient as fitting the pseudo-model. For fixed λ, $L_{\max}(\beta, \lambda)$ is maximized with respect to β by the least-squares estimate, $\hat{\beta}(\lambda)$, when $h(y,\lambda)$ is fit to $h(f(x,\beta),\lambda)$. By the 'max' notation, $L_{\max}(\lambda) = \max_\beta L_{\max}(\beta,\lambda) = L_{\max}(\hat{\beta}(\lambda),\lambda)$. When λ is scalar, as in the modified power family with a fixed shift, $L_{\max}(\lambda)$ can be maximized over λ by a univariate numerical routine or graphically. This is the method that Box and Cox (1964) advocated for their 'transform-the-response' model. However, Box and Cox considered only linear models, and since they did not transform the model, they could fit $h(y, \lambda)$ to $f(x, \beta) = x^T \beta$ by linear least squares. When the model $f(x, \beta)$ is also transformed, nonlinear least squares must be used. Then fitting the pseudo-model (4.17) is easier than repeatedly fitting $h(y, \lambda)$ to $h(f(x, \beta), \lambda)$. We had originally suggested the Box–Cox method (Carroll and Ruppert, 1984a) before discovering the pseudo-model approach.

We have never encountered irremediable convergence problems when fitting the pseudo-model to data, but the model is quite nonlinear in λ, and there may be convergence difficulties for poorly behaved data. In such cases one could try the Box–Cox method of maximizing $L_{\max}(\lambda)$, but if the convergence difficulties are due to the poor behavior of $f(x, \beta)$ then this method probably will not help.

Convergence problems are usually overcome by evaluating the sum of squares over a large grid on the parameter space in order to find a good starting value. Also, poor convergence may indicate either an inadequate or an overparameterized model, $f(x, \beta)$.

One advantage of plotting $L_{\max}(\lambda)$ is that this provides a confidence interval for λ (see below).

4.3.2 Standard errors

There are at least six approaches to the calculation of standard errors. These will be listed and given the following italicized names.

(1) *Fisher information* – Invert the 'observed' Fisher information matrix for the joint estimation of β, λ, and σ, to obtain the large-sample covariance matrix of $\hat{\beta}$, $\hat{\lambda}$, and $\hat{\sigma}$ (see section 7.1). We borrow the terms 'observed' and 'expected' information from Efron and Hinkley (1978); see below.

(2) *Concentrated Fisher information* – Invert the negative Hessian of $L_{\max}(\beta, \lambda)$ evaluated at $\hat{\beta}$ and $\hat{\lambda}$, acting as though $L_{\max}(\beta, \lambda)$ were a loglikelihood for β and λ, to obtain a large-sample covariance matrix for $\hat{\beta}$ and $\hat{\lambda}$. $L_{\max}(\beta, \lambda)$ is often called the concentrated likelihood, though this term could be applied to $L_{\max}(\lambda)$.

(3) *Pseudo-model* – Use the standard errors of $\hat{\beta}$ and $\hat{\lambda}$ when the pseudo-model (4.17) is fit by nonlinear least squares. As will be discussed further, this method does not consistently estimate the standard deviation of $\hat{\lambda}$.

(4) *Fixed λ* – Use the standard errors for $\hat{\beta}$ when $y^{(\lambda)}$ is fit to $f^{(\lambda)}(x, \beta)$ by least squares, λ being fixed at $\hat{\lambda}$ and treated as known.

(5) *M-estimation* – If the error distribution F is not normal, then the normal-theory estimators can still be used, but they will not be true maximum-likelihood estimators. The Fisher information methods will not be consistent for the variance–covariance matrix of $(\hat{\beta}, \hat{\lambda})$. However, the maximum-likelihood estimators do estimate reasonable transformation parameters (Hernandez and Johnson, 1980). Moreover, the MLEs are M-estimators so their variance–covariance matrix can be estimated using the theory of M-estimation in section 7.1. As discussed in that section, when the data can be transformed to normality and homoscedasticity, then using the theory of M-estimation is equivalent asymptotically to using the Fisher information. The advantage of the M-estimation approach is that the standard errors are consistent under very general conditions. F need not be assumed normal and the transformed response need not have a constant variance.

(6) *Bootstrap* – Use Efron's (1979) bootstrap. The bootstrap is described in Efron and Tibshirani (1986), Efron (1982), and in section 2.5.

Each approach gives a large-sample covariance matrix either for $\hat{\beta}$ or for $\hat{\beta}$ and $\hat{\lambda}$ jointly. Confidence intervals for single parameters or confidence regions for sets of parameters can then be constructed using the joint asymptotic normality of $\hat{\beta}$ and $\hat{\lambda}$.

Carroll and Ruppert (1984a, 1987) have studied the first four approaches in some detail.

The following points will help when choosing one of these methods.

(a) The standard large-sample theory for maximum likelihood is not strictly applicable to the Fisher and concentrated Fisher information methods unless the distribution F of the errors is normal. Since the y are assumed positive, this is possible only in the special case where $\lambda = 0$. However, if F is close to normal then we expect the information methods to be only slightly biased.

(b) By a theorem in Patefield (1977), the inverse concentrated information matrix is equal to the covariance matrix of $(\hat{\beta}, \hat{\lambda})$ obtained by inverting the full information matrix and taking the appropriate submatrix. The concentrated information, of course, does not estimate the standard error of $\hat{\sigma}$ or the covariance of $\hat{\sigma}$ with $\hat{\beta}$ or $\hat{\lambda}$. However, statistical inference about σ is often unimportant; typically a point estimate of σ suffices.

(c) The pseudo-model and fixed-λ methods are consistent for the limiting covariance matrix of $\hat{\beta}$ as $N \to \infty$ and $\sigma \to 0$. Such 'small σ' asymptotics have been used by Bickel and Doksum (1981) and others in the study of transformation models where 'fixed-σ' asymptotics are too complicated to offer much insight. 'Small-σ' asymptotics have often proved to be good approximations to finite-sample results when checked by Monte Carlo methods. Moreover, in many data sets, especially from engineering and the physical sciences, σ does seem small in the sense that the model fits the data very well.

(d) The pseudo-model estimates are inconsistent for the variance of $\hat{\lambda}$ or the covariance of $\hat{\lambda}$ with the $\hat{\beta}$, with either σ fixed or σ tending to 0 as N tends to ∞. The fixed-λ method does not provide a standard error for $\hat{\lambda}$. The pseudo-model and fixed-λ methods for estimating the covariance matrix of $\hat{\beta}$ can be supplemented with a consistent estimate of the variance of $\hat{\lambda}$. This estimate is

$$-\left(\frac{\partial^2}{\partial \lambda^2} L_{\max}(\hat{\lambda}) \right)^{-1}$$

The second derivative is easily calculated numerically. Let ε be a small positive constant, say $\varepsilon = 0.01$. Then evaluate $L_{\max}(\lambda)$ at $\hat{\lambda}, \hat{\lambda} - \varepsilon$, and $\hat{\lambda} + \varepsilon$. The approximate second derivative is

$$[L_{\max}(\hat{\lambda} + \varepsilon) + L_{\max}(\hat{\lambda} - \varepsilon) - 2L_{\max}(\hat{\lambda})]/\varepsilon^2$$

For any given value of λ, $L_{\max}(\lambda)$ is obtained as $L_{\max}(\lambda) = \{ -(N/2)\log[N^{-1}\mathrm{SSR}(\lambda)] - N/2 \}$ where $\mathrm{SSR}(\lambda)$ is the regression sum of squares when fitting the pseudo-model, i.e., minimizing (4.16) over β, with λ fixed.

(e) The M-estimation and bootstrap methods do not depend upon normality and will produce consistent estimates of the standard errors under rather general conditions. Moreover as previously mentioned, the M-estimation method will produce consistent estimates of variances even when the transformation does not achieve homoscedasticity. The standard errors derived from M-estimation theory are similar to jackknife standard errors. Wu (1986) advocates the jackknife for its robustness to heteroscedasticity. The ordinary bootstrap is not robust to heteroscedasticity, though a modified bootstrap could be made robust; see the discussions of Wu (1986). As Carroll and Ruppert (1986) point out in the discussion of Wu (1986), robustness to heteroscedasticity is not a major concern; severe heteroscedasticity should be treated by weighting. However, this extra robustness of the M-estimation method is at least of some benefit. An advantage of the bootstrap is its ability to estimate bias. Now we will discuss implementation of these methods.

Fisher information Let $\theta = (\beta, \lambda, \sigma)$ be the parameter vector. Let $l_i = l_i(\theta)$ be the loglikelihood for the *i*th observation so that $L = \sum_{i=1}^{N} l_i$; see equation (4.12). Then the expected Fisher information matrix is

$$I = E\left(\sum_{i=1}^{N} (\nabla l_i)(\nabla l_i)^T \right) = -E\left(\sum_{i=1}^{N} \nabla^2 l_i \right) \tag{4.19}$$

where ∇l_i is the gradient of l_i with respect to θ, and $\nabla^2 l_i = (\nabla \nabla^T) l_i$ is the Hessian matrix. The observed Fisher information matrix is

$$\hat{I} = -\sum_{i=1}^{N} \nabla^2 l_i(\hat{\theta}) \tag{4.20}$$

If L is maximized by Newton's method then $\hat{I}$ is computed as the negative Hessian of L. Another algorithm for maximizing L is the Fisher method of scoring. This is essentially Newton's method but with $\hat{I}$ replaced by

$$\hat{I}_S = \sum_{i=1}^{N} [\nabla l_i(\hat{\theta})][\nabla l_i(\hat{\theta})]^T \tag{4.21}$$

We will call $\hat{I}_S$ the scoring-method observed information matrix.

Concentrated Fisher information It follows from (4.14) that the gradient of $L_{\max}(\beta, \lambda)$ evaluated at $(\hat{\beta}, \hat{\lambda})$ is

$$\nabla L_{\max}(\hat{\beta}, \hat{\lambda}) = -[1/\hat{\sigma}^2(\hat{\beta}, \hat{\lambda})] \sum_{i=1}^{N} e_i(\nabla e_i)$$

where $e_i = e(x_i, \hat{\beta}, \hat{\lambda})$ is defined by the formula preceding (4.17), and ∇e_i is the gradient of e_i with respect to β and λ. Since $(\hat{\beta}, \hat{\lambda})$ maximizes $L_{\max}(\beta, \lambda)$,

$$\sum_{i=1}^{N} e_i(\nabla e_i) = 0$$

and therefore the Hessian of $L_{\max}(\hat{\beta}, \hat{\lambda})$ is

$$\nabla^2 L_{\max}(\hat{\beta}, \hat{\lambda}) = -[\hat{\sigma}^{-2}(\hat{\beta}, \hat{\lambda})]$$
$$\times \left(\sum_{i=1}^{N} (\nabla e_i)(\nabla e_i)^T + \sum_{i=1}^{N} e_i(\nabla^2 e_i) \right) \tag{4.22}$$

Note that the Fisher method of scoring is not applicable to $L_{max}(\beta, \lambda)$, since that function is not expressed as a sum of components for $y_1, \ldots, y_N$.

To simplify (4.22) note that $\log(\dot{y})$ is an average of all the values of $\log(y_i)$ so it will be approximately equal to its expectation. If we ignore the randomness of $\dot{y}$ then only the lower-right corner of $\nabla^2 e_i$ is random, i.e., depends on y_i. Therefore only the lower-right corner of $E(e_i \nabla^2 e_i)$ is nonzero and

$$\sum_{i=1}^{N} (e_i \nabla^2 e_i) \simeq \begin{bmatrix} 0 & 0 \\ 0 & H \end{bmatrix}$$

where $H = -\sum_{i=1}^{N} e_i [\partial^2/\partial \lambda^2) e_i]$. Therefore

$$s^2 \left(\sum_{i=1}^{N} (\nabla e_i)(\nabla e_i)^T + \begin{bmatrix} 0 & 0 \\ 0 & H \end{bmatrix} \right)^{-1} \qquad (4.23)$$

can be used to estimate the variance–covariance matrix of $(\hat{\beta}, \hat{\lambda})$.

Pseudo-model The estimated variance–covariance matrix when fitting the pseudo-model $D_i = e(x_i, \beta, \lambda)$ is

$$s^2 \left(\sum_{i=1}^{N} (\nabla e_i)(\nabla e_i)^T \right)^{-1}$$

which is similar to (4.23) but without the term involving H. This explains why the pseudo-model estimates are not consistent for the standard deviation of $\hat{\lambda}$. One could call estimator (4.23) the corrected pseudo-model estimate.

Fixed λ The estimated variance–covariance matrix of $\hat{\beta}$ using nonlinear regression with λ fixed at $\hat{\lambda}$ is

$$\tilde{s}^2 \left(\sum_{i=1}^{N} [(\partial/\partial \beta)h(f(x_i, \hat{\beta}), \hat{\lambda})][(\partial/\partial \beta)h(f(x_i, \hat{\beta}), \hat{\lambda})]^T \right)^{-1}$$

Here $\tilde{s}^2$ is $N/(N - p)\hat{\sigma}^2$, which differs from (4.18) only in the degrees of freedom being $(N - p)$ not $(N - p - 1)$ since λ is not estimated. Except for this difference, the information and fixed-λ methods would give the same estimates of the variance–covariance matrix of $\hat{\beta}$ if the Fisher information matrix were block-diagonal (with $\hat{\beta}$ asymptotically independent of $(\hat{\lambda}, \hat{\sigma})^T$). This is precisely what happens under small-σ asymptotics; see Carroll and Ruppert (1984a).

M-estimation Huber's (1967) asymptotic theory of *M*-estimation is outlined in section 7.1. In that section, the observations $y_1, \ldots, y_N$ are independent with distributions $F_1, \ldots, F_N$. The parameter θ is estimated by solving

$$\sum_{i=1}^{N} \Psi_i(y_i, \hat{\theta}) = 0$$

where $\Psi_1, \ldots, \Psi_N$ are 'estimating functions'. Here we let Ψ_i be the Fisher score function assuming that the data can be transformed to normality and constant variance. The F_i are the *true* distributions of the data and are *not* based on this assumption. The point is that Huber's theory of *M*-estimation tells us the effects of model misspecification on asymptotic distributions. The estimated variances of $\hat{\beta}$ and $\hat{\lambda}$ from this theory are consistent under misspecification.

In the examples to follow, we use the gradient of $L_{\max}(\beta, \lambda)$ as the estimating function Ψ_i. The estimated variance–covariance matrix of $(\hat{\beta}, \hat{\lambda})$ is $\hat{B}^{-1}\hat{A}(\hat{B}^{-1})^{T}$, where $\hat{A}$ and $\hat{B}$ are given by (7.11) and (7.12). The derivatives of $\dot{\Psi}_i$ in (7.12) are computed numerically.

Bootstrap The bootstrap is applied to weighted least squares in section 2.5, where possible refinements are also mentioned. Here we outline the bootstrap for transformation models. Again we let $\theta = (\beta, \lambda, \sigma)$. Let

$$\varepsilon_i = h(y_i, \hat{\lambda}) - h(f(x_i, \hat{\beta}), \hat{\lambda})$$

be the *i*th residual. Take a bootstrap sample $\varepsilon_1^*, \ldots, \varepsilon_N^*$ by sampling *with* replacement from $\{\varepsilon_1, \ldots, \varepsilon_N\}$ and let

$$y_i^* = h^{-1}[h(f(x_i, \hat{\beta}), \hat{\lambda}) + \varepsilon_i^*]$$

Let $\hat{\theta}^*$ be the MLE of θ from the bootstrap sample $\{y_1^*, \ldots, y_N^*\}$. This process is repeated N_B times and the variance–covariance matrix of $\hat{\theta}$ is estimated by

$$N_B^{-1} \sum_{i=1}^{N} (\hat{\theta}_i^* - \bar{\theta}^*)(\hat{\theta}_i^* - \bar{\theta}^*)^{T} \qquad (4.24)$$

where $\bar{\theta}_* = N_B^{-1}\sum_{i=1}^{N_B} \hat{\theta}_i^*$. The bootstrap estimate of bias is $(\bar{\theta}^* - \hat{\theta})$.

There is a variant of the bootstrap that we will call the parametric bootstrap. This technique uses a parametric estimate of the distribution F of the errors. Let K be the largest absolute residual. Let Φ_θ be the normal distribution with mean equal to 0 and standard deviation equal to $\hat{\sigma}$. Then let $\hat{F}_P$ be $\Phi_{\hat{\sigma}}$ truncated below at $-K$ and above at K. The parametric bootstrap uses an independent, identically distributed sample of size N from $\hat{F}_P$ in place of $\varepsilon_1^*, \ldots, \varepsilon_N^*$.

4.3.3 Confidence intervals

Confidence regions for the components of β and/or λ or some subset of these parameters can be constructed using the asymptotic normality of the estimators. Confidence regions of this type will be called Wald regions since they are based on work of Wald (1943); see section 7.2. The agreement between the true and nominal coverage probabilities of Wald regions can be poor if the parameterization is sufficiently 'nonlinear'. The accuracy of Wald regions for nonlinear models is best understood for nonlinear regression, where the application of differential geometry by Bates and Watts (1980) sheds much light on the problems of nonlinearity. Bates and Watts (1980) classify the effects of nonlinearity into two types, intrinsic curvature and parametric-effects curvature. The latter, but not the former, can be removed by suitable nonlinear reparameterization. However, there are no general methods to search for such reparameterizations and, when possible, it seems better to construct likelihood ratio confidence regions which are invariant to reparameterizations and are affected only by intrinsic nonlinearity.

The method for constructing likelihood ratio confidence regions is well known but will be briefly described here. Let $L(\gamma, \varphi)$ be a loglikelihood depending upon two vector parameters, γ and φ, with γ of dimension r. As discussed in section 7.2, to test $H_0: \gamma = \gamma_0$ one uses

$$LR = 2[L(\hat{\gamma}, \hat{\varphi}) - L(\gamma_0, \hat{\varphi}(\gamma_0))]$$

where $(\hat{\gamma}, \hat{\varphi})$ is the unrestricted MLE of (γ, φ) and $\hat{\varphi}(\gamma_0)$ is the MLE of φ under the constraint that $\gamma = \gamma_0$. The null hypothesis is rejected if $LR > \chi_{1-\alpha}^2(r)$, the $(1 - \alpha)$ quantile of the chi-squared distribution with r degrees of freedom. Therefore, a $100(1 - \alpha)\%$ confidence region for γ is

$$\{\gamma : L(\gamma, \hat{\varphi}(\gamma)) > L(\hat{\gamma}, \hat{\varphi}) - \tfrac{1}{2}\chi_{1-\alpha}^2(r)\}$$

As an application of this technique, the likelihood ratio confidence interval for λ in the TBS model is

$$\{\lambda : L_{\max}(\lambda) > L_{\max}(\hat{\lambda}) - \tfrac{1}{2}\chi_{1-\alpha}^2(1)\}$$

Confidence intervals for the components of β can be constructed in an analogous manner. Likelihood ratio intervals are a greater computational burden than Wald intervals. This is their major disadvantage. Likelihood ratio intervals can be expected to see more widespread use in nonlinear models as advances are made in hardware and, perhaps more importantly, in statistical software. Wald confidence regions are always ellipsoids and can be described by a quadratic form, but likelihood ratio confidence regions can be of arbitrary form and are, in general, difficult to describe for three or more parameters.

4.3.4 *Conditional inference – treating λ as fixed*

It follows from (c) in the discussion of standard errors (subsection 4.3.2) that the standard errors of $\hat{\beta}$ calculated by the fixed-λ method agree under small-σ asymptotics with the standard error calculated by methods acknowledging that λ is estimated. This is an important point. Moreover, the Monte Carlo results suggest that, even when σ is not small, treating λ as known will not lead us astray when we make inferences about β (Carroll and Ruppert, 1984a).

In the case of the Box–Cox 'transform-the-response only' model, the situation is quite different, as Bickel and Doksum (1981) have demonstrated. In the Box–Cox model $\hat{\beta}$ may be orders of magnitude more variable when λ is estimated than when it is known and fixed. The reason is that as λ varies there will be drastic changes in the response $h(y, \lambda)$ but not in the model $f(x, \beta)$. This has led to considerable controversy; see Hinkley and Runger (1984) and the discussion to that paper. One point of contention is whether β is a 'physically meaningful parameter' when λ is not known exactly. In contrast, when one transforms the response and the model, then β is a physically meaningful parameter even without knowledge of λ; $f(x, \beta)$ is the conditional median of y given x.

In summary, the controversy over whether λ can be treated as fixed is not relevant to the transform-both-sides model.

4.3.5 *Alternative estimators*

The maximum-likelihood estimator attempts to transform to normality and homoscedasticity. As has been pointed out, transformation to exact normality is not possible when y is positive except for certain values of λ. A more modest and attainable goal is to transform to either symmetry or homoscedasticity, or perhaps both.

In this section we introduce an estimator $\hat{\lambda}_{sk}$ transforming to symmetry and an estimator $\hat{\lambda}_{het}$ transforming to homoscedasticity. By comparing them we can see whether the goals of symmetry and homoscedasticity after transformation are compatible.

Transformations to symmetry have been studied by Hinkley (1975) and Taylor (1985). Here we introduce them in the setting of transform both sides. Let $T(\hat{F}_N)$ be a symmetry measure applied to an empirical distribution function $\hat{F}_N$. Thus $T(\hat{F}_N)$ should be 0 for a perfectly symmetric sample, positive for right-skewed data, and negative for left-skewed data. The standard skewness coefficient, i.e. the sample third moment divided by s^3, is a typical choice of T. Another possibility is

$$\{[\hat{F}_N^{-1}(\tfrac{3}{4}) - \hat{F}_N^{-1}(\tfrac{1}{2})]/[\hat{F}_N^{-1}(\tfrac{1}{2}) - \hat{F}_N^{-1}(\tfrac{1}{4})]\} - 1$$

which depends only on the sample quartiles and is insensitive to outliers.

For fixed λ, regress $h(y_i, \lambda)$ on $h(f(x_i, \beta), \lambda)$, let $\hat{\beta}(\lambda)$ be the least-squares estimate, and let $\hat{F}_N(\lambda)$ be the empirical distribution of the residuals. Let $T(\lambda)$ be $T[\hat{F}_N(\lambda)]$. Then $\hat{\lambda}_{sk}$ is defined by the estimating equation

$$T(\hat{\lambda}_{sk}) = 0 \qquad\qquad (4.25)$$

Let $\hat{\beta}_{sk} = \hat{\beta}(\hat{\lambda}_{sk})$. Clearly, $(\hat{\beta}_{sk}, \hat{\lambda}_{sk})$ is an M-estimate, the defining equations being the least-squares 'normal equations' and (4.25).

The estimator $\hat{\lambda}_{het}$ is based on a test for heteroscedasticity. There are many methods of testing for heteroscedasticity in regression. These are discussed in Chapter 3. Suppose H is a test statistic for heteroscedasticity that depends on the absolute residuals and predicted values. Suppose as well that $H = 0$ corresponds to no heteroscedasticity, positive H indicates that the variance is an increasing function of the mean, and negative H indicates that the variance is a decreasing function of the mean. For example, H could be the Pearson correlation between the squared residuals and the logarithms of the predicted values; this is the 'score test' for heteroscedasticity in Chapter 3. One might prefer the Spearman rank correlation because of its robustness. In that case, it is irrelevant whether we use squared or absolute residuals or whether we use predicted values or their logarithms; the Spearman rank correlation is invariant to rank-preserving, i.e., monotonically increasing, transformations of either variable.

Now for fixed λ let $H(\lambda)$ be the value assumed by H when $h(y_i, \lambda)$ is regressed on $h(f(x_i, \beta), \lambda)$ and H is applied to the absolute residuals and predicted values. Then $\hat{\lambda}_{\text{het}}$ is defined by the equation

$$H(\hat{\lambda}_{\text{het}}) = 0 \qquad (4.26)$$

and $\hat{\beta}_{\text{het}}$ is $\hat{\beta}(\hat{\lambda}_{\text{het}})$.

Aldershof and Ruppert have begun a study of $\hat{\lambda}_{\text{sk}}$ and $\hat{\lambda}_{\text{het}}$. They also consider a transformation to both symmetry and heteroscedasticity which they call $\hat{\lambda}_{\text{hetsk}}$. The latter is defined by a linear combination of equations (4.25) and (4.26). The weights in this linear combination are estimated to minimize the asymptotic variance of $\hat{\lambda}_{\text{hetsk}}$. Since this research has not been completed, we are unable to give further details. In section 4.5, the values of these estimates on a real data set will be compared with the maximum-likelihood estimate.

4.4 Inference about the dependent variable

One typically thinks of regression analysis as modeling the conditional mean of y given x, but regression analysis actually includes more. Whether or not it includes a transformation, a regression model expresses the entire conditional distribution of the dependent variable y as a function of the independent variables x. Consider first an ordinary multiple linear regression model

$$y = x^{\mathrm{T}}\beta + \sigma\varepsilon$$

where ε has distribution Φ, the standard normal distribution function. Often statistical inference concentrates on β, which of course only describes the conditional mean (and median) of y given x. The normality assumption plus knowledge of σ actually tells us the entire conditional distribution. For example

$$P(y_i \leqslant y \mid x_i) = \Phi[(y - x_i^{\mathrm{T}}\beta)/\sigma] \qquad (4.27)$$

and the pth quantile of y_i given x_i is

$$q_p(y_i \mid x_i) = x_i^{\mathrm{T}}\beta + \sigma\Phi^{-1}(p) \qquad (4.28)$$

Conditional probabilities or quantiles for y_i can be estimated by substituting $\hat{\beta}$ and $\hat{\sigma}$ into (4.27) and (4.28).

Now consider the TBS model

$$h(y, \lambda) = h(f(x, \beta), \lambda) + \varepsilon$$

or re-expressing this equation as a model for y itself

$$y = h^{-1}\{[h(f(x, \beta), \lambda) + \varepsilon], \lambda\}$$

where ε has distribution F. Until now we have assumed that $F(x) = \Phi(x/\sigma)$, but we know this assumption cannot be strictly correct for families of transformations whose range is not the entire real line. When y is positive this would preclude the modified power transformations except when $\lambda = 0$. Therefore, we will only assume that F is close to normal but with bounded support, perhaps a truncated normal distribution. We will also assume that h is monotonically increasing.

The pth quantile of y given x is

$$q_p(y|x) = h^{-1}\{[h(f(x, \beta), \lambda) + F^{-1}(p)], \lambda\} \qquad (4.29)$$

A possible estimator of $q_p(y|x)$ is

$$\hat{q}_p(y|x) = h^{-1}\{[h(f(x, \hat{\beta}), \hat{\lambda}) + \hat{\sigma}\Phi^{-1}(p)], \hat{\lambda}\} \qquad (4.30)$$

This estimator is parametric in that $\hat{\sigma}\Phi^{-1}(p)$ is used as an estimate of the pth quantile of the error distribution. Because of concern over the normality assumption, we might replace $\hat{\sigma}\Phi^{-1}(p)$ by the pth quantile of the residuals, $\hat{F}_N^{-1}(p)$, where $\hat{F}_N$ is the empirical distribution function of the residuals. Then a nonparametric estimator of $\hat{q}_p(y|x)$ is

$$\tilde{q}_p(y|x) = h^{-1}\{[h(f(x, \hat{\beta}), \hat{\lambda}) + \hat{F}_N^{-1}(p)], \hat{\lambda}\} \qquad (4.31)$$

Given a new x, a $100(1 - \alpha)\%$ prediction interval for y given x is

$$(\hat{q}_{\alpha/2}(y/x) - q_{1-\alpha/2}(y/x)) \qquad (4.32)$$

and of course the nonparametric $\tilde{q}$ could be used instead of $\hat{q}$.

Formula (4.32) ignores estimation errors for β, λ, and σ. Using (4.32) is reasonable only when the sample size is sufficiently large. Correcting for estimation error by modifying (4.32) would complicate the calculation of the prediction interval, but is probably necessary for small to moderately large data sets. In example 4.1 of section 4.5, the sample size is 27 and the prediction limits are surprisingly sensitive to λ.

In other areas of statistics, e.g., time series analysis (Box and

Jenkins, 1976), sample sizes are large and it is standard practice to treat the parameters as known when setting prediction limits. However, in calibration and prediction problems, the variability of $\hat{\beta}$ and $\hat{\sigma}$ are accounted for. The conditional mean of y given x is

$$E(y|x) = \int h^{-1}\{[h(f(x,\beta),\lambda) + \varepsilon], \lambda\} \, dF(\varepsilon)$$

To estimate $E(y|x)$ we can substitute estimates for β and λ, but if h^{-1} is not defined on the entire real line, as is usually the case, e.g., for power transformations, we cannot simply substitute $\Phi(\cdot/\hat{\sigma})$ as an estimator of F unless we truncate the range of integration. A parametric estimator of $E(y|x)$ is

$$\hat{E}(y|x) = [\Phi(\max\{r/\hat{\sigma}\}) - \Phi(\min\{r/\hat{\sigma}\})]^{-1}$$
$$\times \int_{\min\{r\}}^{\max\{r\}} h^{-1}\{[h(f(x,\hat{\beta}),\hat{\lambda}) + \hat{\sigma}\varepsilon], \hat{\lambda}\} \, d\Phi(\varepsilon)$$

$$(4.33)$$

where $\min\{r\}$ and $\max\{r\}$ are, respectively, the smallest and largest residuals. A nonparametric alternative to (4.33) is the 'smearing estimator' of Duan (1983). Into the definition of $E(y|x)$ this estimator substitutes $(\hat{\beta}, \hat{\lambda})$ for (β, λ) and $\hat{F}_N$ for F. Specifically, the smearing estimator is

$$\tilde{E}(y|x) = N^{-1} \sum_{i=1}^{N} h^{-1}\{[h(f(x,\hat{\beta}),\hat{\lambda}) + r_i], \hat{\lambda}\} \qquad (4.34)$$

In an unpublished Monte Carlo study, we found $\hat{E}(y|x)$ and $\tilde{E}(y|x)$ have similar performance when the F is the normal distribution truncated at $\pm 3\sigma$. Standard errors for either $\hat{E}(y|x)$ or $\tilde{E}(y|x)$ could be obtained using either the delta method or the bootstrap, but the performance of such techniques in this context has not been adequately studied.

The estimation of an expectation is inherently nonrobust. $\hat{E}(y|x)$ may perform badly if no member of the family $h(y, \lambda)$ can transform to nearly normal, constant-variance, errors.

$\tilde{E}(y|x)$ is based on the assumptions that, for some $\lambda, h(y, \lambda)$ will transform to errors that are identically distributed, and that $\hat{\lambda}$ consistently estimates this λ. We would not expect that the normal-theory MLE will be consistent, though its asymptotic bias may be acceptable in practice. However, the alternative estimators discussed at the end of section 4.3 will be consistent.

Retransforming to estimate the conditional median was introduced by Carroll and Ruppert (1981a) for the Box–Cox model where the regression function is not transformed. In this context, Taylor (1986) studied the smearing estimator and an estimator based on approximating $E(y/x)$ by a Taylor series.

4.5 Examples

Example 4.1   Skeena River sockeye salmon

Background   When managing a fishery, one must model the relationship between the size of the annual spawning stock and its production of new catchable-sized fish, called recruits or returns. There are several theoretical models relating recruits (R) and spawners (S). These are derived from simple assumptions about factors influencing the survival of juvenile fish. All spawner–recruit models known to us are deterministic, i.e., R is nonrandom given S, though S itself can depend upon stochastic variables. If the biological and physical factors affecting fish survival were constant from year to year, then a deterministic model would be realistic since abundance of fish makes the law of large numbers applicable. However, for most fish stocks these factors are far from constant. There has been little work on stochastic models for recruitment, probably because the mechanisms causing survival rates to vary are not well understood. It is common practice to take a deterministic model relating R and S and to assume multiplicative lognormal errors. The transform-both-sides approach allows us to test this assumption, and to model the errors empirically when the assumption seems unwarranted.

Ricker (1954) derived the theoretical deterministic model

$$R = \beta_1 S \exp(-\beta_2 S) = f_{RK}(S, \beta) \qquad \beta_2 \geqslant 0 \qquad (4.35)$$

In this model $f_{RK}(S, \beta)$ tends to 0 as S goes to 0, as would be expected in any realistic model. Moreover, $f_{RK}(S, \beta)$ has a maximum at β_2^{-1}, provided β_2 is strictly positive, and $f_{RK}(S, \beta)$ tends to 0 as S goes to ∞. The biological interpretation of this behavior is that as the number of juveniles increases, increased competition and predation affect the survival rate so drastically that the absolute number of juveniles reaching maturity decreases. A second model was derived by

Beverton and Holt (1957), namely

$$R = 1/(\beta_1 + \beta_2/S) = f_{BH}(S, \beta) \qquad \beta_1 \geqslant 0 \text{ and } \beta_2 \geqslant 0 \quad (4.36)$$

(Interestingly, the same function is the Michaelis–Menten model of enzyme kinetics.) The Beverton–Holt model also has the characteristic that R tends to 0 as S tends to 0, but R increases asymptotically

Table 4.1   *Skeena River sockeye salmon data. Units are thousands of fish*

Year	Spawners	Recruits
1940	963	2215
1941	572	1334
1942	305	800
1943	272	438
1944	824	3071
1945	940	957
1946	486	934
1947	307	971
1948	1066	2257
1949	480	1451
1950	393	686
1951	176	127
1952	237	700
1953	700	1381
1954	511	1393
1955	87	363
1956	370	668
1957	448	2067
1958	819	644
1959	799	1747
1960	273	744
1961	936	1087
1962	558	1335
1963	597	1981
1964	848	627
1965	619	1099
1966	397	1532
1967	616	2086

(Source: Ricker and Smith, 1975)

to $1/\beta_1$ as S tends to ∞. It is natural to think of $1/\beta_1$ as the carrying capacity of the environment, the maximum number of recruits that the available space, food, and other resources can support. When fit to the same data set, the Ricker and Beverton–Holt models are often similar over the range of spawner values in the data, despite qualitatively different behavior as the number of spawners increases to infinity (see the discussion below).

Ricker and Smith (1975) give numbers of spawners and recruits from 1940 until 1967 for the Skeena River sockeye salmon stock. Their data are given in Table 4.1. A rockslide occurred in 1951 and severely reduced the number of recruits; we will not use this observation in the present analysis, though it will be shown in some graphs. Also, it will be used in Chapter 6 to demonstrate the behavior of robust estimators. Ricker and Smith (1975) mention that the slide

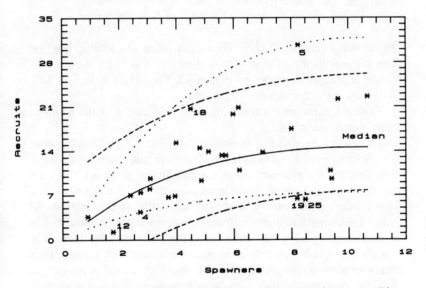

Figure 4.1   *Skeena River sockeye salmon. Plot of recruits and spawners. The full curve is the estimated median recruitment. The broken curves are the estimated 5th and 95th percentiles using nonlinear least squares without a transformation. The dotted curves are the estimated 5th and 95th percentiles by transform both sides. All estimates are based on the Ricker model. Selected cases are indicated. Case 12 is included in the graph but was not used during estimation. Spawners and recruits are in units of 100 000 fish.*

affected 1952 as well, but our analysis did not find case 13 to be anomalous. Year 1955 (case 16) has an outlier in S since the spawning stock that year came from the 1951 recruitment. However, case 16 has a recruitment value that is not unusual (conditional on the low value of S) and this case was retained throughout the analysis.

Figure 4.1 is a scatterplot of recruits and spawners. The estimated median recruitment and the prediction limits in the figure will be explained later in this section. R is nearly a linear function of S. This does not contradict the Ricker or Beverton–Holt models since they are somewhat linear over the range of S in the data. A linear relationship between spawners and recruits would be a suitable model for some purposes but not for studying optimal management policies, since it suggests that recruitment could be increased indefinitely by reducing fishing and allowing more spawners. Furthermore, the scatterplot shows clear heteroscedasticity with the variance increasing either as a function of S or the mean.

Parameter estimation   The TBS model using the Ricker function and the modified power transformation $y^{(\lambda)}$ was fit by the pseudo-model technique described in section 4.3. The MLEs were $\hat{\beta}_1 = 3.78$, $\hat{\beta}_2 = 9.54 \times 10^{-4}$, and $\hat{\lambda} = -0.20$.

Table 4.2 contains the standard errors of $\hat{\beta}$ and $\hat{\lambda}$ by the methods discussed in section 4.3.

The standard errors of $\hat{\beta}$ by ordinary nonlinear regression (without a transformation) are also included. These are calculated in two ways. The first are the estimates from a least-squares program; these are inconsistent because of the heteroscedasticity. The second method uses Huber's (1967) theory of M-estimation discussed in sections 4.3 and 7.1. This method does produce consistent variance estimates. The standard errors from the least squares program seem too large. The standard errors from the M-estimation theory are slightly larger than the standard errors obtained from the TBS model, and are an indication that estimation without a transformation is inefficient.

The fixed-λ method was used with $\lambda = -0.2$ (the MLE) and the commonly used log-transformation, $\lambda = 0$. The fixed-λ, pseudo-model, and concentrated Fisher information methods produced similar standard errors for $\hat{\beta}$, which is in agreement with the asymptotic theory. The standard error of $\hat{\lambda}$ is quite different for the pseudo-model method than for the Fisher information method, but we know that the pseudo-model method is not consistent for the standard error of $\hat{\lambda}$.

Table 4.2 *Skeena River sockeye salmon. Standard errors of $\hat{\beta}$ and $\hat{\lambda}$ by ordinary nonlinear regression and TBS regression. Observation 12 (year 1951) deleted*

	Estimator		
Method	$\hat{\beta}_1$	$\hat{\beta}_2$	$\hat{\lambda}$
Nonlinear regression			
Usual standard errors	1.0*	(3.6×10^{-4})*	–
M-estimation theory	0.77	3.4×10^{-4}	–
Transform both sides			
Fisher information	0.66	3.0×10^{-4}	0.35
Pseudo-model	0.71	3.3×10^{-4}	0.62*
Fixed λ ($\lambda = -0.2$)	0.69	3.1×10^{-4}	–
($\lambda = 0$)	0.75	3.2×10^{-4}	–
M-estimation theory	0.58	3.3×10^{-4}	0.27
Numerical second derivative of $L_{\max}$ at $\lambda = -0.2$ using $\varepsilon = 0.1$	–	–	0.36
Bootstrap	0.72	3.09×10^{-4}	0.32
Parametric bootstrap	0.77	3.09×10^{-4}	0.38

*Not a consistent method of estimating the true standard deviation.

The second derivative of $L_{\max}$ was calculated numerically, as described in section 4.3, using $\varepsilon = 0.1$. The resulting standard error of $\hat{\lambda}$ is in reasonable agreement with the other consistent methods.

The large-sample confidence interval for λ by the Fisher information method is

$$-0.2 \pm (1.96)(0.35) = (-0.89, 0.52)$$

and the same interval using more robust *M*-estimation theory is

$$-0.2 \pm (1.96)(0.27) = (-0.73, 0.33)$$

The likelihood ratio confidence interval for λ is constructed graphically in Figure 4.2, a plot of $2L_{\max}$. We can see that $L_{\max}(\lambda)$ obtains its maximum at the MLE, -0.2. The 0.95 quantile of the chi-squared distribution with one degree of freedom is 3.84. The horizontal line through $y = 2L_{\max}(-0.2) - 3.84$ intersects the graph of $2L_{\max}$ at $\lambda = -0.89$ and $\lambda = 0.52$ giving the confidence interval $(-0.89, 0.52)$. The three methods of constructing confidence intervals are in close

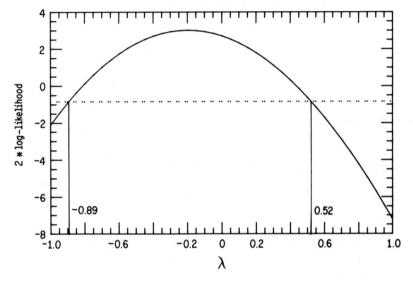

Figure 4.2   *Skeena River sockeye salmon. Profile loglikelihood and 95% confidence interval for λ. Case 12 excluded from the likelihood. A constant has been subtracted from the loglikelihood.*

agreement. The confidence intervals are rather long, but two cases of special interest, namely $\lambda = 1$, indicating no transformation, and $\lambda = -1$, which linearizes the Beverton–Holt model, can be ruled out. The log-transformation which linearizes the Ricker model and is consistent with the common assumption of multiplicative lognormal errors is supported by the data, being well within the intervals. Most statisticians would prefer to use the log-transformation rather than the MLE. The log-transformation has several advantages including ease of interpretation and an easily computed estimate of the expected recruitment, $\exp[f_{RK}(x, \hat{\beta}) + \hat{\sigma}^2/2]$.

If the Beverton–Holt model is used instead of the Ricker model, the point estimates and confidence intervals for λ are virtually unchanged.

Residual analysis   To explore these data further, the estimated median recruitments

$$\text{Median}_i = f_{\text{RK}}(S_i, \hat{\beta}) \qquad (4.37)$$

and the untransformed residuals from the median recruitments

$$e_i = R_i - \text{Median}_i \qquad (4.38)$$

were computed. The estimated medians are shown in Figure 4.1 as the full curve. Figure 4.3 is a plot of the residuals against the medians and suggests that the assumption of a constant coefficient of variation would be tenable, at least as a rough approximation. From section 4.2 we know that a constant coefficient of variation implies that $\lambda = 0$ is the variance-stabilizing transformation. This is in reasonable agreement with the MLE, $\hat{\lambda} = -0.2$, but not the transformation to a zero value of the score test for heteroscedasticity, $\hat{\lambda}_{\text{het}} = -0.86$ (see below).

Figure 4.4 is a normal probability plot, that is, a plot of the residual probits against the residuals, where the ith residual probit is

$$\Phi^{-1}[(e_i - \tfrac{3}{8})/(N + \tfrac{1}{4})]$$

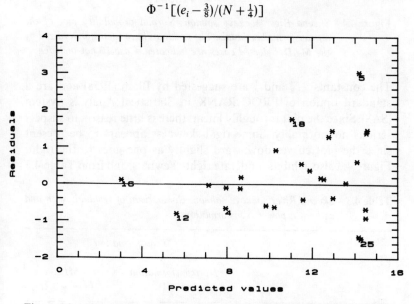

Figure 4.3   *Skeena River sockeye salmon. Residuals and predicted values from the maximum-likelihood estimate expressed in the original, not the transformed, scale. The residuals have been standardized by the median absolute deviation (MAD). Selected cases are indicated. Case 12 is included in the graph but was not used in the estimation.*

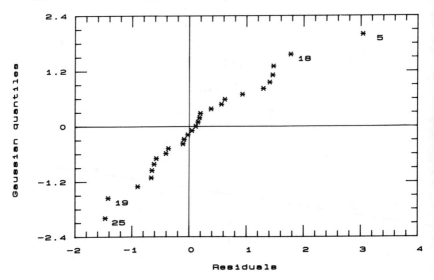

Figure 4.4   *Skeena River sockeye salmon. Normal probability plot of the untransformed residuals. The residuals have been standardized by the MAD. Selected cases are indicated. Case 12 not included.*

The constants $-\frac{3}{8}$ and $\frac{1}{4}$ are suggested by Blom (1958) and are a standard option of PROC RANK in Statistical Analysis System (SAS). Since the plot is roughly linear, there is little reason to suspect serious nonnormality. Some right-skewness appears to be present since the plot curves downward slightly as one goes to the right. Figure 4.3 also exhibits moderate right-skewness, and from Table 4.3

Table 4.3   *Skeena River sockeye salmon. Comparison of residuals with and without a power transformation*

	Transformation	
	No transformation ($\lambda = 1$)	*MLE* ($\lambda = -0.2$)
Skewness coefficient	0.46	-0.46
Kurtosis coefficient	0.34	-0.51
Spearman correlation between absolute residuals and medians	0.56 ($p = 0.0014$)	0.28 ($p = 0.15$)

we see that the skewness coefficient of the residuals is 0.46. This degree of skewness would not be of concern except that it is also accompanied by heteroscedasticity. This normal probability plot should not be analyzed too closely since the residuals are not identically distributed; we know that they exhibit substantial heteroscedasticity. Normal probability plots are only designed to check whether identically distributed random variables are normal, and it is not clear what is revealed when heteroscedastic variables are plotted.

To see the effects of the transformation, let

$$e_i(\hat{\lambda}) = R_i^{(\hat{\lambda})} - \text{Median}_i^{(\hat{\lambda})}$$

Figure 4.5 is a plot of $e_i(\hat{\lambda})$ against Median_i. A quick look may suggest heteroscedasticity, but the data are bunched on the right side of the plot, and in fact there is little indication of heteroscedasticity. The Spearman rank correlation given in Table 4.3 between $|e_i(\hat{\lambda})|$ and Median_i is only 0.28 ($p = 0.15$), while the Spearman correlation between $|e_i|$ and Median_i is 0.56 ($p = 0.0014$). The p-values are calculated assuming an i.i.d (independent, identically distributed) bivariate sample and are not strictly valid here, but are included for

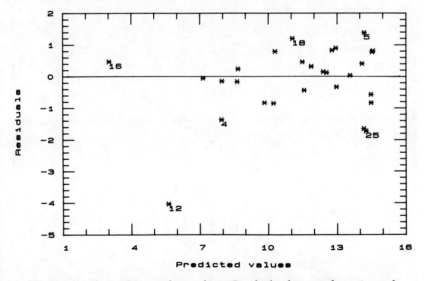

Figure 4.5 *Skeena River sockeye salmon. Residuals after transformation and predicted values on the untransformed scale. Residuals have been standardized by the MAD. Selected cases are indicated. Case 12 is shown on the graph but was not used for estimation.*

descriptive purposes. Case 16 with its low spawner value and small residual certainly increases the appearance of heteroscedasticity in Figures 4.3 and 4.5. The Spearman rank correlations, of course, are not too sensitive to this case. The transformation has substantially reduced the association between the absolute residuals and the predicted values.

A comparison of Figures 4.3 and 4.4 with Figures 4.5 and 4.6 suggests that the transformation has reduced the slight right-skewness in the original data, perhaps inducing left-skewness. In Table 4.3 we see that the skewness coefficient of the residuals has changed sign from 0.46 to -0.46. The transformation changes the number of positive residuals from 15 out of 27 to 12 out of 27. The MLE overtransforms from right- to left-skewness, because variance stabilization requires a stronger transformation than symmetrization (see below).

The kurtosis of the residuals is also reduced by transformation, from 0.34 to -0.51. The positive kurtosis before transformation is probably due to the heteroscedasticity, since a scale mixture of

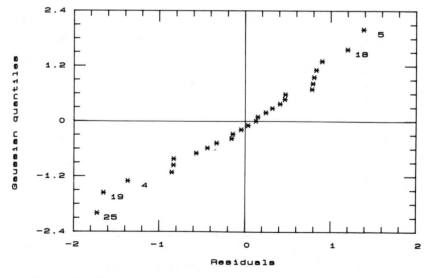

Figure 4.6   *Skeena River sockeye salmon. Normal probability plot of residuals after transformation. Residuals have been standardized by the MAD. Selected cases are indicated. Case 12 excluded.*

normal distributions is more heavy-tailed than the normal distribution itself. Therefore, the change in the kurtosis of the residuals is most likely due to variance stabilization rather than to a change in distributional shape.

The alternative estimators of λ discussed in section 4.3 were also tried. The transformation to a zero value of the residual third-moment skewness coefficient is $\hat{\lambda}_{sk} = 0.45$, but a far more severe transformation, $\hat{\lambda}_{het} = -0.86$, is needed to achieve a zero value of the score test for heteroscedasticity.

These estimates and the skewness coefficients in Table 4.3 show that the MLE, $\hat{\lambda} = -0.2$, overtransforms from slight right- to slight left-skewness in order to remove most, but not all, of the heteroscedasticity. A more severe transformation to homoscedasticity would induce more severe left-skewness. The fact that the transform to zero skewness ($\hat{\lambda}_{sk} = 0.45$) is midway between no transformation ($\lambda = 1$) and the MLE ($\hat{\lambda} = -0.2$) is no doubt the reason that the MLE changes the sign but not the magnitude of the skewness coefficient.

Note that $\hat{\lambda}_{sk}$ and $\hat{\lambda}_{het}$ are on opposite extremes of the 95% likelihood ratio confidence interval, $(-0.89, 0.52)$, and outside the more robust M-estimation confidence interval, $(-0.73, 0.33)$.

The minimum-asymptotic-variance combined skewness–heteroscedasticity estimate is $\hat{\lambda}_{hetsk} = -0.26$, which is close to the MLE.

Estimating recruitment The most notable improvement by the transform-both-sides model over not transforming is not in the estimation of median or mean recruitment but rather in estimating recruitment variability. We calculated the nonlinear least-squares fit when the data are not transformed, and the estimates were $\tilde{\beta}_1 = 3.95$ and $\tilde{\beta}_2 = 8.53 \times 10^{-4}$, which are not far from the MLEs based on TBS. However, the estimated conditional distribution of R given S is considerably different for the TBS nonlinear regression model than for the ordinary nonlinear regression model. Using the nonlinear regression model, 90% prediction intervals for R given S are

$$f_{RK}(S, \tilde{\beta}) \pm 1.65\tilde{\sigma} \tag{4.39}$$

where $\tilde{\sigma} = \sqrt{(MSE)} = 566$. Using the TBS model, the 90% prediction intervals are

$$[f_{RK}(S, \hat{\beta})^{-0.2} \pm 1.65\hat{\sigma}]^{-5} \tag{4.40}$$

where $\hat{\sigma}$ is the root-mean-square error after power-transforming both sides using $\lambda = -0.2$. The prediction limits are plotted in Figure 4.1. The limits given by (4.39) are shown as broken curves. They are of course parallel and do not fit the data well. Of the top eight spawner values, three of the eight corresponding recruitment values are outside the broken curves, while of the eight smallest spawner values, none of the eight recruitments come close to the broken curves.

The limits given by (4.40) and shown as dotted curves exhibit skewness and heteroscedasticity similar to that in the data. However, like (4.39), formula (4.40) underestimates the recruitment variability for large values of S. Of the eight cases with S greater than 750, two are still outside the limits and one is on the boundary. Using $\hat{\lambda} < -0.2$ would not entirely correct this, since the skewness and the heteroscedasticity would change together. Although for large values of S the upper prediction limits would increase, the lower limits would remain about the same (see below).

One might expect prediction limits based on the common assumption of multiplicative lognormal errors to be similar to those based on $\lambda = -0.2$, since the log-transformation fits the data almost as well as the MLE transformation. However, this is not the case. The upper prediction limits depend heavily on λ. In Table 4.4 we show the 90% prediction limits for various values of $\lambda : \lambda = -0.86 = \hat{\lambda}_{het}$; $\lambda = -0.3$, which is the robust 'redescending' estimate (Carroll and Ruppert, 1987); $\lambda = -0.2$, which is the MLE; $\lambda = 0$, which is the log-

Table 4.4   *Skeena River sockeye salmon. 90% prediction intervals for recruitment. Ricker spawner–recruit model. The three spawner values are the lowest and highest observed values and a typical value*

	Spawners (thousands of fish)		
λ	87	511	1066
-0.86	244, 394	660, 4091	744, 8569
-0.3	197, 495	635, 2564	752, 3331
-0.2	185, 524	625, 2473	749, 3142
0	152, 601	600, 2336	738, 2865
0.3	87, 747	548, 2219	709, 2634
0.5	28, 872	498, 2170	678, 2531
1.0	-621, 1226	263, 2110	534, 2381

transformation; $\lambda = 0.3$, which is the MLE if the rockslide year is included (see Carroll and Ruppert, 1987); $\lambda = 0.5$, which is the upper limit of the confidence intervals for λ and near $\hat{\lambda}_{sk}$; and $\lambda = 1.0$, corresponding to no transformation. The prediction intervals are given for spawner numbers (in thousands) equal to 87, the lowest observed value; 1066, the highest observed value; and 511, an intermediate value. Notice that the upper prediction limits vary substantially, even between $\lambda = -0.2$ and $\lambda = 0$. The lower prediction limits are stable, unless $\lambda \geqslant 0.5$, but such λ values are outside the 95% confidence interval for λ.

More research is needed on prediction intervals based on transformation models. Unless the sample size is far greater than 27, the effects of estimation error in the parameters needs to be taken into account, but the best way to do this is uncertain.

Choosing between the Ricker and Beverton–Holt models   So far we have considered the regression model $f(x, \beta)$ fixed when using the transform-both-sides methodology, but TBS is equally useful when choosing between alternative models. In this example, both the Ricker and Beverton–Holt models should be considered. The log-transformation linearizes the Ricker model to

$$\log(R/S) = \alpha_1 + \alpha_2 S \qquad (4.41)$$

where $\alpha_1 = \log(\beta_1)$ and $\alpha_2 = -\beta_2$. The Beverton–Holt model is linearized by the inverse transformation to

$$1/R = \beta_1 + \beta_2(1/S) \qquad (4.42)$$

Multiplying through by S in (4.42) gives another linearized model

$$S/R = \beta_2 + \beta_1 S \qquad (4.43)$$

One might be tempted to choose between the Beverton–Holt and Ricker models by comparing (4.41) to either (4.42) or (4.43). A much abused measure of fit is the squared multiple correlation R^2. The values of R^2 for models (4.41), (4.42), and (4.43) are 0.25, 0.53, and 0.24. Based upon these values one might naïvely select model (4.42). However, models should be compared either using the same transformation, or fitting each model using the MLE transformation for that model. Moreover, R^2 is not necessarily a relevant statistic for model selection.

The Beverton–Holt and Ricker models are both special cases of the model

$$R = [\exp(-\beta_2 S)]/(\beta_3 + \beta_1/S) \qquad (4.44)$$

since (4.44) reduces to the Beverton–Holt model when $\beta_2 = 0$ and to the Ricker model when $\beta_3 = 0$. The MLE of λ is -0.236 for model (4.44) and -0.167 for the Beverton–Holt model. The maximum loglikelihood is -165.1 for the Beverton–Holt model and -165.0 for both the Ricker model and (4.44). Clearly the Ricker and Beverton–Holt models fit these data equally well and the more general model (4.44) provides no improvement. A comparison based on linearizing the models and using R^2 is misleading. In some examples, e.g., the Atlantic menhaden data studied by Carroll and Ruppert (1984a), linearization seems to favor the model whose linearizing transformation is closest to the MLE. Here model (4.42) is apparently selected for a different reason; R is more dependent upon S than is R/S, the so-called production ratio or average number of recruits per spawner. In fact R/S is nearly constant for these data. In models (4.41) and (4.43), the response is a function of R/S, so these models have small R^2 values.

Example 4.2 Population 'A' sockeye salmon

This is another example of sockeye salmon data. The source of the data has not granted permission to reveal the name of the river. These data were called 'population A' in Ruppert and Carroll (1985). Figure 4.7 is a scatterplot of the recruits and spawners. The 90% prediction limits and the estimated median recruitment based on TBS applied to the Beverton–Holt model are shown as dotted curves and a full curve, respectively. Notice that, compared to the Skeena River population, this population has a less pronounced relationship between R and S, and recruitment is more highly skewed and is perhaps less heteroscedastic. However, since recruitment appears almost independent of the size of the spawning population, even if the variance is a function of the mean (or median) recruitment, one would expect less heterogeneity of variance simply because the mean does not vary much. A transformation was needed for the Skeena River data primarily to reduce heteroscedasticity; the data were nearly normal both before and after the transformation. In the present example, we will see that the transformation is needed primarily to remove skewness. The heteroscedasticity is little affected by transformation.

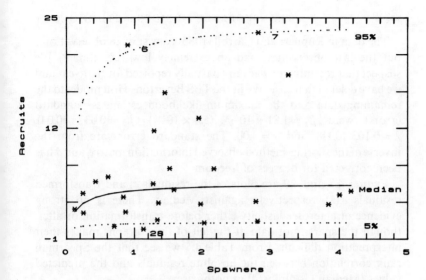

Figure 4.7 Population 'A' sockeye salmon. Recruits and spawners. The dotted curves are the estimated 5th and 95th percentiles. The full curve is the estimated median. All estimates use transform both sides. Case 28 is included in the graph but not used for estimation. Spawners and recruits are in units of 10 000 fish.

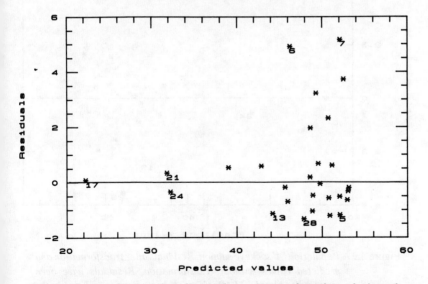

Figure 4.8 Population 'A' sockeye salmon. Untransformed residuals and predicted values. Residuals have been standardized by the MAD. Case 28 shown but not used in the estimation.

The data in Ruppert and Carroll (1985) consist of 28 observations. but the last observation had an extremely low recruitment. We suspect that recruitment was only partially reported for that year and we have deleted that case. We fit the TBS Beverton–Holt model to the remaining data and the maximum-likelihood estimates (standard errors) were $\hat{\beta}_1 = 1.81 \times 10^{-5}$ (5.4×10^{-5}), $\hat{\beta}_2 = 0.032$ (0.04), $\hat{\lambda} = 0.103$ (0.18), and $\hat{\sigma} = 3.00$. The standard errors are from the inverse of the scoring-method observed information matrix, and $\hat{\sigma}$ has been corrected for degrees of freedom.

Figures 4.8 and 4.9 are plots of the residuals e_i and transformed residuals $e_i(\hat{\lambda})$, respectively, against Median$_i$. There is no strong evidence of heteroscedasticity either before transformation or after, though the sparsity of data at the left of these graphs makes their interpretation difficult. From Table 4.5 we see that the Spearman rank correlation between the absolute residuals and the predicted values (Median$_i$) is almost 0 in both cases.

It should be emphasized that Table 4.5 used residuals from $\hat{\beta} =$ MLE even when $\lambda = 1$, i.e., we always transformed using the MLE of λ when estimating β. If $\hat{\beta}$ were the ordinary, nonlinear least-squares

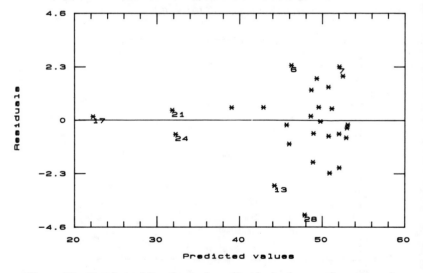

Figure 4.9   *Population 'A' sockeye salmon. Residuals after transformation and predicted values without transformation. Residuals have been standardized by the MAD. Case 28 shown but not included in the estimation.*

Table 4.5   *Population 'A' sockeye salmon. Comparison of residuals with and without a power transformation. Beverton–Holt model*

	Transformation	
	No. transformation ($\lambda = 1$)	MLE ($\lambda = 0.103$)
Skewness coefficient	1.46	− 0.13
Kurtosis coefficient	1.27	− 0.23
Spearman correlation	− 0.06	− 0.04
between absolute	($p = 0.78$)	($p = 0.86$)
residuals and medians		

estimate, then the results in Table 4.5 for $\lambda = 1$ would have been rather different: the skewness, kurtosis, and Spearman correlation would have been 1.3, 0.7, and 0.44, respectively. The large changes in the kurtosis and the Spearman correlation are surprising since the fitted curve is only moderately affected by the change in $\hat{\beta}$; the fitted curve using $\lambda = 1$ has a somewhat higher asymptote (as $S \to \infty$) than the fitted curve using $\lambda = 0.103$. This example shows that the p-value for the Spearman correlation between the absolute residuals and the medians can, at best, be used only for descriptive purposes since the effects of estimating β have not been taken into account.

Figures 4.10 and 4.11 are normal probability plots of the residuals e_i and transformed residuals $e_i(\hat{\lambda})$, respectively. The extreme right-skewness before transformation and the near-symmetry after are evident. From Table 4.5 we see that the skewness of the residuals was reduced from 1.46 to − 0.13 by the transformation. The kurtosis was also brought closer to 0, the value under normality.

When β_2 is 0, then the Beverton–Holt model reduces to the constant-mean model. Since $\hat{\beta}_2$ is considerably smaller than its standard error, the constant-mean model is plausible. The loglikelihood is − 12.36 for the constant-mean model and − 11.85 for the Beverton–Holt model. The constant-mean model seems quite adequate over the range of spawner levels in the data, though it should not be extrapolated to low spawner levels.

The loglikelihood for the Ricker model is − 12.51, which is lower than the constant-mean model with fewer parameters. The Ricker

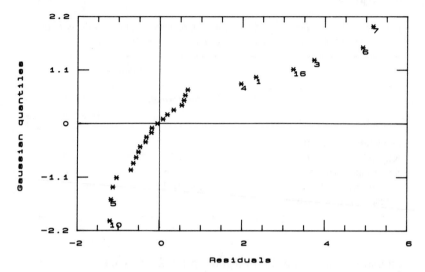

Figure 4.10 *Population 'A' sockeye salmon. Normal probability plot of residuals without transformation. Residuals have been standardized by the MAD. Case 28 excluded.*

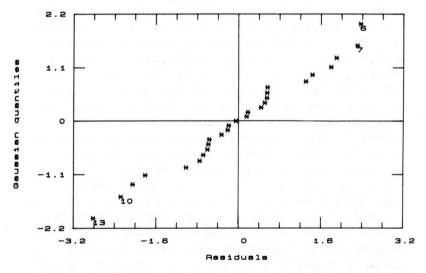

Figure 4.11 *Population 'A' sockeye salmon. Normal probability plot of the residuals after transformation. Residuals have been standardized by the MAD. Case 28 excluded.*

model does not include the constant-mean model as a special case. The fitted Ricker function reaches a peak at slightly less that 20 000 spawners and decreases substantially by 40 000 spawners, but this behavior does not seem consistent with the data.

For these reasons we chose the Beverton–Holt model.

Example 4.3   Bacterial clearance by the lung

The lungs of a healthy animal have several defense mechanisms for killing inhaled bacteria, for example, ingestion by macrophages. At the beginning of this chapter we mentioned an experimental technique to study these mechanisms. Mice are placed in an exposure chamber containing an aerosol flow of bacteria. At selected times after exposure, the mice are sacrificed and the number of bacteria entering the lung and the number of viable bacteria present in the lung at sacrifice are determined. In the experiment we are about to analyze, the mice were exposed for 30 min and then sacrificed at 0, 4, or 24 h after exposure. The mice were each given one of four treatments:

O   Control.
A   Administration of the antibiotic ampicillin shortly before exposure.
V   Infection by a virus several days prior to exposure.
VA Ampicillin and virus infection.

The experiment was a $2^2 \times 3$ full factorial with factors A = ampicillin, V = virus, and T = time of sacrifice, and with six replicates per cell. The response was the number of viable bacteria in the lung at the time of sacrifice expressed as a percentage of the viable bacteria that entered the lung during exposure in the aerosol chamber. A conventional approach to this example might be a factorial analysis, perhaps with a transformation of the response, but a theoretical model where the parameters have interesting biological interpretations is preferable.

The data are given in Table 4.6. The data exhibit extreme heteroscedasticity and, for some cells, marked right-skewness. In mice not infected by virus prior to exposure, the number of bacteria in the lung decreased rapidly. Virus-infected mice had a net increase of bacteria in their lungs, especially if they had not been given the antibiotic. Since the response varies over eight orders of magnitude, a re-expression of the data, say a log-transformation, is necessary even

Table 4.6   *Bacterial clearance data. The response is the ratio (expressed as a percentage) of viable bacteria in the lung at the time of sacrifice to viable bacteria in the lung at the end of exposure. Case numbers are given in parentheses for observations of special interest in later chapters*

Group	0	4	24
		Time of sacrifice (h)	
O Control	103.0	26.4	0 (no. 49)
	92.5	52.7	1.5
	72.1	45.5	1.7
	41.4 (no. 4)	63.1	0.6
	63.7	29.1	2.0
	74.2	27.8	1.5
A Ampicillin-	103.0	29.6	1.1
treated	92.6	25.9	0.5
	54.6	15.4	0.2 (no. 57)
	75.5	48.7 (no. 34)	0.8
	42.8 (no. 11)	14.1	2.1
	77.1	14.2	2.0
V Virus-	135.1	258.0	3365
infected	154.0	139.0	197400
	157.0	96.5	172000
	83.9	519.6	211 (no. 64)
	74.8	453.0	8110
	107.0	368.0	2798
VA Virus-infected	128.0 (no. 19)	36.2	565
and ampicillin-	89.2	23.4	439
treated	95.2	22.1	1719
	90.1	28.3	38.1
	95.7	43.0	303
	96.4	112.4 (no. 48)	3.3 (no. 72)

for informal graphical analysis. Some of the response values were 0 (to the accuracy of the experiment), and 0.05 was added to all values of the response before the log-transformation, and later other transformations, were taken. In the notation of section 4.1, a fixed shift $\mu = -0.05$ was used.

When the cell means of the log-transformed data are plotted against time, one sees a roughly linear change in the means over time within the groups O and V. The change in log bacterial counts is somewhat curvilinear for the groups A and VA, probably indicating that the bacterial death rate decreased over time as the antibiotic was eliminated. It is reasonable to hypothesize that the net growth rate ('birth' minus death rates) of bacteria in the O and V groups is constant over time. As a simple starting model we will assume that within each non-ampicillin-treated group, the bacteria population changes at a rate that is constant over time, but that this rate depends upon whether or not the mouse had been virus-infected. For ampicillin-treated mice we assume that the bacterial death rate is piecewise constant with a change at 4 h after exposure. This leads to the model

$$y_{itj} = \exp\{\alpha_i + \beta_i(t + 0.25) + \Delta_i(t - 4)I[t > 4]\} \qquad (4.45)$$

where y_{itj} is the response of the jth mouse among those in the ith treatment group that were sacrificed at time t, and $I[t > 4]$ is the indicator that t exceeds 4. Here $i = O, A, V,$ or $VA, t = 0, 4,$ or $24,$ and $j = 1, \ldots, 6$. The parameter β_i is the net bacterial growth rate, i.e., 'birth' rate minus death rate, in the ith group of mice during the period $t = 0$ to $t = 4$. The parameter Δ_i is the change in this growth rate after 4 h. It is assumed that $\Delta_1 = \Delta_3 = 0$, because the first and third groups did not receive the ampicillin treatment. Since t is the time between exposure and sacrifice, we add 0.25 to t because the bacteria are in the lungs about 15 min (0.25 h) on average during exposure.

Model (4.45) is deterministic, but in reality there are many sources of variability. The experimental procedure allows several sources of statistical sampling error. For example, to estimate numbers of bacteria, researchers dilute the lung homogenate and culture a sample of the dilution on a petri dish. Each viable bacteria, or cluster of bacteria, forms a visible colony. Also, the bacteria are radiolabeled, and the number that entered the lung, viable or not, is estimated by putting the homogenate in a scintillation counter. Moreover, the mice are not homogenous and β_i presents only an average net growth rate for the ith group; individual mice within a group will have their own net growth rates that vary about β_i.

It seems impossible to model all these sources of variation. Instead, we will fit the TBS model with the modified power transformations

Table 4.7   *Bacterial clearance data. Means and standard deviations of the residuals. The residuals are standardized by the MAD (median absolute deviation). The model for the median is equation (4.45). The TBS model was used with the shifted power transformation family; the shift was fixed at $\mu = -0.05$ (i.e., 0.05 was added to the response and the model)*

		Time of sacrifice (h)		
Group		0	4	24
O Control	Mean	−0.03	0.22	−0.04
	SD	0.56	0.67	3.23
A Ampicillin-treated	Mean	0.09	0.09	0
	SD	0.58	0.92	1.78
V Virus-infected	Mean	0	0.19	0.05
	SD	0.59	1.12	3.67
VA Virus-infected and ampicillin-treated	Mean	0.08	0.09	0.09
	SD	0.23	1.07	3.99

applied to equation (4.45), i.e., the model

$$(y_{itj} + 0.05)^{(\lambda)} = (0.05 + \exp\{\alpha_i + \beta_i(t + 0.25) + \Delta_i(t - 4)I[t > 4]\})^{(\lambda)} + \varepsilon_{itj} \qquad (4.46)$$

The maximum-likelihood estimates are given later in Table 5.1.

The within-cell means and standard deviations of the residuals are given in Table 4.7. The small cell means indicate that model (4.45) for the median responses is adequate. Clearly the variance increases with time. A plot (not included here) of the residuals indicates that they are roughly symmetric. The power transformation is only successful in transforming to symmetry. This is understandable since the within-cell variances appear to depend heavily upon time as well as on the mean response. In the next chapter we introduce a model that allows both transformations and weighting, thus combining the 'weight-only' model of Chapter 3 and the 'transform-only' model of this chapter. As we will see, this 'transform and weight' model, with the weights a function of time, fits the bacterial clearance data considerably better than 'transform-only' or 'weight-only' models.

CHAPTER 5

Combining transformations and weighting

5.1 Introduction

In the last chapter we saw that a single transformation may induce both normally distributed and homoscedastic errors. However, it should not be surprising that for some data sets the transformation that induces normality will not stabilize the variance. What is surprising is how often a single transformation *will* do both reasonably well! Nevertheless, in some situations we need a transformation and weighting in order that the errors are reasonably close to normal with a constant variance. This is particularly true when the variance depends on a covariate. In the bacterial clearance example of the last chapter, we saw that the variance depends heavily on the time of sacrifice as well as the mean response, and the transformation could only remove the second source of heteroscedasticity.

Moreover, for some data it is unclear whether a transformation or weighting is preferable even though probably both are not needed. This is often true when the variance is a monotonic function of the mean, for example the Skeena River sockeye salmon data in the last chapter. Then we can test which is preferable, weighting or transformation, by using a model that includes both possibilities. This is not to say that the test will necessarily be conclusive; often weighting or transformation are equally satisfactory.

In this chapter we study the model

$$h(y_i, \lambda) = h(f(x_i, \beta), \lambda) + \sigma g(\mu_i(\beta), z_i, \theta)\varepsilon_i \tag{5.1}$$

where y_i is the response, $h(y, \lambda)$ is a member of a parametric family of transformations as in Chapter 4, $g(\mu_i(\beta), z_i, \theta)$ is a function expressing the heteroscedasticity of the data, $\mu_i(\beta) = f(x_i, \beta), z_i$ is a known variable, perhaps a function of x_i, θ is a variance parameter, and σ is a scaling parameter. We allow g to depend on the mean response as well as the covariate z_i, but in practive g will usually not depend on $\mu_i(\beta)$ since that type of heteroscedasticity can be removed by the transformation. An exception might be where the transformation to symmetry differs greatly from the transformation to variance independent of the mean. Then the transformation to symmetry will be selected by maximum likelihood if the function g properly models the dependence of the variance on the mean.

Often for some value of θ, say θ_0, $g(\mu_i(\beta), z_i, \theta_0)$ is constant and (5.1) reduces to the standard TBS model of Chapter 4. Often, as well, $h(y, \lambda_0)$ is the identity function for some λ_0, so (5.1) includes the heteroscedastic models of Chapters 2 and 3 as special cases. By testing H_0: $\lambda = \lambda_0$, we can see if a heteroscedastic model in the original response y adequately describes the data. Similarly, if we test H_0: $\theta = \theta_0$ then we are testing whether a transformation alone can bring about normally distributed and homoscedastic errors.

Much of the theory developed in the last three chapters for weighting or transformation alone will be relevant when combining them. Estimation by maximum likelihood is straightforward in theory and not difficult to implement using modern software. Inference for the untransformed response is possible by extending the estimators, e.g., the smearing estimate, of section 4.4.

One new difficulty is that the estimated transformation parameter $\hat{\lambda}$ and the estimated variance parameters $\hat{\theta}$ can be highly correlated. This means that they cannot be estimated jointly with the same degree of accuracy as when only one is present in the model. In Ruppert and Carroll (1985), the Skeena River and population 'A' salmon stocks are analyzed with the TBS model plus a variance function that is a power of the independent variable, spawners. Neither the transformation parameter nor the power parameter can be estimated well, unless the other is treated as fixed and known. The bacteria data of Example 4.3 in section 4.5 are an exception. The transformation and variance parameters there have different and almost independent effects (see section 5.3).

5.2 Parameter estimation

In this section we introduce the maximum-likelihood estimators of β, σ, λ, and θ in model (5.1). For certain special cases of practical importance, the loglikelihood can be put in a form that is very convenient for computation and the MLEs can be found using standard nonlinear regression software. We will emphasize these computational techniques.

As mentioned in the last section, generally the variance model g will not depend on the mean response $\mu_i(\beta)$. But when it does, the maximum-likelihood estimates will have the robustness problem mentioned in Chapter 3. The MLE of β will use sample information about the variance. This increases the efficiency of $\hat{\beta}$ when the variance is correctly specified but biases $\hat{\beta}$ under misspecification. The generalized least-squares estimate does not have this bias. For this reason, one might re-estimate β by GLS after $\hat{\lambda}$ and $\hat{\theta}$ have been found by maximum likelihood, treating $\hat{\lambda}$ and $\hat{\theta}$ as fixed. This can be done by the methods of Chapter 3.

By model (5.1), the conditional density of y_i given x_i is

$$f(y_i \mid x_i, \beta, \lambda, \theta, \sigma) = [2\pi\sigma^2 g^2(\mu_i(\beta), z_i, \theta)]^{-1/2}$$
$$\exp[-r_i^2(\beta, \lambda, \theta)/2\sigma^2]J_i(\lambda)$$

where, as in Chapter 4, $J_i(\lambda)$ is the Jacobian of the transformation $\varepsilon_i \to h(y_i, \lambda)$ and here we let

$$r_i(\beta, \lambda, \theta) = [h(y_i, \lambda) - h(f(x_i, \beta), \lambda)]/g(\mu_i(\beta), z_i, \theta)$$

The loglikelihood for the data is, apart from an additive constant,

$$L(\beta, \lambda, \theta, \sigma) = -N\log(\sigma) - \sum_{i=1}^{N} r_i^2(\beta, \lambda, \theta)/(2\sigma^2)$$

$$- \sum_{i=1}^{N} \log[g(\mu_i(\beta), z_i, \theta)] + \sum_{i=1}^{N} \log[J_i(\lambda)]$$

Given β, λ, and θ, the MLE of σ^2 is

$$\hat{\sigma}^2(\beta, \lambda, \theta) = N^{-1} \sum_{i=1}^{N} r_i^2(\beta, \lambda, \theta)$$

and the loglikelihood maximized over σ is

$$L_{\max}(\beta, \lambda, \theta) = L(\beta, \lambda, \theta, \hat{\sigma}(\beta, \lambda, \theta))$$
$$= -(N/2)\log[\hat{\sigma}^2(\beta, \lambda, \theta)]$$
$$- N\log[\dot{g}(\beta, \theta)] + N\log[\dot{J}(\lambda)] - N/2$$

where $\dot{g}(\beta, \theta)$ and $\dot{J}(\lambda)$ are the geometric means of $\{g(\mu_i(\beta), z_i, \theta)\}$ and $\{J_i(\lambda)\}$. Simple algebra shows that

$$L_{\max}(\beta, \lambda, \theta) = -(N/2)\log\left(N^{-1} \sum_{i=1}^{N} [\dot{g}(\beta, \theta)r_i(\beta, \lambda, \theta)/\dot{J}(\lambda)]^2\right) - N/2$$

Therefore, the MLE minimizes

$$\sum_{i=1}^{N} [\dot{g}(\beta, \theta) r_i(\beta, \lambda, \theta) / \dot{J}(\lambda)]^2 \tag{5.2}$$

Expression (5.2) can be minimized by nonlinear software that allows sufficient flexibility, e.g., 'do loops', in the programming statements defining the model. To use standard software, the 'pseudo-model' method described in section 4.3 can be used. The dummy variable D_i, which is identically zero, is fit to the regression model

$$D_i = [\dot{g}(\beta, \lambda)/\dot{J}(\lambda)] r_i(\beta, \lambda, \theta) \tag{5.3}$$

As explained in section 4.3, the purpose of the dummy response is that the residuals from (5.3) are the same as those from (5.1), but in (5.3) the response does not depend on parameters. Therefore, (5.3) can be fit by standard software.

Since $\dot{g}(\beta, \lambda)$ and $\dot{J}(\lambda)$ are functions of all the data, the right-hand side of (5.3) depends on all cases, not just the ith, and fitting (5.3) typically requires multiple passes through the data at each iteration of the algorithm that minimizes the sum of squares.

In some situations only a single pass is needed. This is the case if $\dot{J}(\lambda)$ is a power function of λ and $\dot{g}(\beta, \lambda)$ is independent of β and a power function of λ. For example, if $h(y, \lambda)$ is the modified power transformation family (4.3) then $\dot{J}(\lambda) = \dot{y}^{\lambda - 1}$ as we noted in Chapter 4. If $g(\mu_i(\beta), z_i, \theta) = |f(x_i, \beta)|^\theta$ then $\dot{g}(\beta, \lambda) = |\dot{f}(\beta)|^\theta$ where $\dot{f}(\beta)$ is the geometric mean of $\{f(x_i, \beta)\}$. Although $\dot{g}$ depends on β, we could fix β at a preliminary estimate. Moreover, $\dot{g}$ will be independent of β when

$$g(\mu_i(\beta), z_i, \theta) = z_i^\theta \tag{5.4}$$

Model (5.4) is often appropriate in practice. Model (5.1) with variance model (5.4) and the Box–Cox modified power transformation family was used in Ruppert and Carroll (1985). When one uses a modified power transformation and model (5.4), then the MLE minimizes

$$(\dot{z}^\theta / \dot{y}^{\lambda - 1})^2 \sum_{i=1}^{N} r_i^2(\beta, \lambda, \theta) \tag{5.5}$$

and this minimization can be easily accomplished by fitting the pseudo-model

$$D_i = (\dot{z}^\theta / \dot{y}^{\lambda - 1}) r_i(\beta, \lambda, \theta) \tag{5.6}$$

The beauty of (5.6) is that $\dot{z}$ and $\dot{y}$ can be computed *a priori* and then the right-hand side of (5.6) essentially depends only on the ith case.

Once the parameters have been estimated, then the conditional distribution of y given x and z can be estimated by modifying the techniques in section 4.4. Recalling that for fixed λ, $h^{-1}(y, \lambda)$ is the inverse of $h(y, \lambda)$ as a function of y, we see that

$$y = h^{-1}\{[h(f(x, \beta), \lambda) + \sigma g(\mu_i(\beta), z_i, \theta)\varepsilon], \lambda\}$$

where ε has a distribution F independent of x and z. It is assumed that F is nearly normal but perhaps with bounded support. Then a suitable estimator of the conditional pth quantile of y is

$$\hat{q}_p(y|x, z) = h^{-1}\{[h(f(x, \hat{\beta}), \hat{\lambda}) + \hat{\sigma} g(\mu_i(\hat{\beta}), z_i, \hat{\theta})\Phi^{-1}(p)], \hat{\lambda}\} \quad (5.7)$$

This estimator is a simple extension of equation (4.30) that accounts for the heteroscedasticity of y even after transformation to $h(y, \lambda)$. As in Chapter 4 we are assuming that h and hence h^{-1} are monotonically increasing. Large-sample prediction intervals for y, given x and z, can be constructed from equation (4.32) with $\hat{q}_p(y|x, z)$ defined by (5.7) replacing $\hat{q}_p(y|x)$ defined by (4.31).

The conditional mean of y given x can be estimated either parametrically by generalizing (4.33) or nonparametrically using the 'smearing estimator'. Define $r_i = r_i(\hat{\beta}, \hat{\lambda}, \hat{\theta})$ to be the residual from the MLE standardized by $g(\mu_i(\hat{\beta}), z_i, \hat{\theta})$. Then the smearing estimate is

$$N^{-1} \sum_{i=1}^{N} h^{-1}\{[h(f(x, \hat{\beta}), \hat{\lambda}) + g(u, \hat{\beta}, \hat{\theta})r_i], \hat{\lambda}\}, \mu = f(x, \hat{\beta})$$

5.3 Examples

Example 5.1   Bacterial clearance in the lung

This is a continuation of Example 4.3 in section 4.5. When we fit model (4.45) to the bacteria data using transform both sides, the residuals became more variable over time (see Table 4.6). The increasing relationship between standard deviation and time has a plausible biological interpretation. Recall that β_i is the net bacterial growth rate in treatment group i. This rate will vary between animals. In fact, it varies greatly. In other experiments (uniformity trials) with controls only (Ruppert *et al.*, 1975) the assays of bacterial densities (total bacteria and viable ones only) were replicated to measure within-animal variation. This is the variation due to measurement error. A variance-components analysis showed that within-animal variability was rather smaller than between-animal variation.

Since β_i varies across animals in treatment group i, we have a nonlinear random-coefficients regression model. Suppose that the TBS model (4.46) holds, except that the β_i may be random quantities. If $\lambda = 0$ (log-transformation) were the true value of the transformation parameter and we ignored the shift $\mu = -0.05$, then the random-coefficients model would be linear. The variance would be a linear function of $(t + 0.25)^2$ with intercept and slope equal to the variances of ε_{itj} and β_i. Thus

$$\sigma_{itj} = \sigma g_A(t, \pi) = \{\sigma^2[\pi^2 + (t + 0.25)^2]\}^{1/2} \qquad (5.8)$$

where

$$(\sigma\pi)^2 = \mathrm{Var}(\varepsilon_{itj}) \qquad \text{for all } i, t, \text{ and } j \qquad (5.9)$$

and

$$\sigma^2 = \mathrm{Var}(\beta_i) \qquad \text{for all } i \qquad (5.10)$$

Here σ_{itj} is the standard deviation of $(y_{itj} - \mu)^{(\lambda)}$. The variance function g is subscripted 'A' to distinguish this model from latter ones.

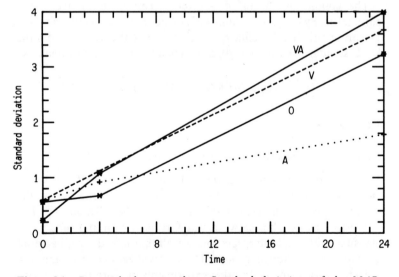

Figure 5.1 *Bacterial clearance data. Standard deviations of the MAD-standardized residuals from the transform-both-sides model. The standard deviations are plotted against time by treatment group. $\mu = -0.05$. Constant-variance model.*

Model (5.8)–(5.10) is certainly plausible, especially since the MLE, $\hat{\lambda} = -0.048$, is so close to the log-transformation. However, there is no compelling reason why the within-animal variation must be multiplicative with a variance independent of time as (5.9) implies. Moreover, it is not clear whether the variance parameters σ and π should be independent of the treatment group as is implied by (5.9) and (5.10). We must examine the data themselves before accepting this model. To see the pattern of variability across time, the within-cell residual variances and standard deviations were plotted against time, separately by treatment group. The first plot, which is not shown, looks roughly quadratic and supports model A. The second plot, Figure 5.1, shows that the standard deviation is nearly a linear function of time, perhaps with the same slope and intercept for all groups. Another possibility, one that will be explored in section 6.5, is that virus-infected animals, groups V and VA, have a different slope than the other mice. In Figure 5.1, groups V and VA look similar. Groups O and A differ at $t = 24$, but this is largely due to an outlier,

Table 5.1   *Bacterial clearance data. Parameter esti-*
mates. The modified power transformation
was used with shift μ fixed at -0.05. Three
variance models were used

	Variance model		
Parameter	Constant variance	Model A	Model B
α_1	4.34	4.35	4.34
α_2	4.29	4.33	4.33
α_3	4.69	4.70	4.70
α_4	4.60	4.65	4.65
β_1	-0.191	-0.185	-0.187
β_2	-0.293	-0.293	-0.292
β_3	0.175	0.185	0.184
β_4	-0.245	-0.248	-0.244
Δ_2	0.133	0.131	0.130
Δ_4	0.314	0.314	0.310
λ	-0.048	-0.069	-0.063
π	$-$	3.51	$-$
θ	$-$	$-$	1.99

no. 49, in group O (see below). For now we will assume a constant intercept and slope. This suggests model B, which can be written

$$\sigma_{itj} = \sigma g_B(t, \theta) = \sigma[\theta + (t + 0.25)] \qquad (5.11)$$

where $\sigma\theta$ and σ are the intercept and slope of the standard deviation as a function of time.

Models A and B are similar when π and θ are close to zero. Both models were fit by minimizing (5.2) using the pseudo-model method on a nonlinear least-squares program. The minimized values of $-2 \times$ (loglikelihood) were 572.2 and 570.4 for models A and B, respectively. The small difference, 1.8, suggests that both models are equally acceptable. The parameter estimates are given in Table 5.1. The estimates of the parameters in the model (4.46) for the median response are remarkably similar, and in both cases $\hat{\lambda}$ is quite close to 0. Within-cell residual standard deviations for models A and B are given in Table 5.2 and seem independent of time and treatment. Either

Table 5.2 *Bacterial clearance data. Standard deviations of the residuals from equation (4.46) using the modified power transformation family with shift μ fixed at -0.05. The residuals were standardized by the median absolute deviation (MAD) before the standard deviations were computed. Variance models A and B were used*

Group			Time of sacrifice (h)		
			0	4	24
O	Control	Model A	1.18	0.91	1.09
		Model B	1.18	1.21	1.09
A	Ampicillin-treated	Model A	1.21	1.26	0.59
		Model B	0.46	0.80	1.13
V	Virus-infected	Model A	1.12	1.46	1.00
		Model B	1.21	1.26	1.58
VA	Virus-infected and ampicillin-treated	Model A	0.47	1.44	1.20
		Model B	0.83	1.01	1.45

model A or B seems acceptable, but we prefer model A because of its biological interpretability.

Model A should be checked further by residual plots. Figure 5.2 is a plot of the residuals against the logarithms of the fitted values. The residuals have been standardized by their median absolute deviation (MAD). The ith fitted value is $f(x_i, \hat{\beta}) - \mu$. No observation is particularly outlying and the residuals are reasonably symmetric. Figure 5.3, a normal probability paper plot, indicates near-normality.

There is a hint of heteroscedasticity remaining. In Figure 5.2 the detached group of observations with very small fitted values consists of the groups O (control) and A (ampicillin) at $t = 24$. Eleven of these twelve cases have absolute residuals less than 1. The other observation is case 49, the zero response.

Overall, the plots show little to concern us, though the slight residual heteroscedasticity might be investigated. Perhaps the variance function should have group-dependent parameters, or at least one parameter for the O and A groups and another parameter for the V and VA groups. We will return to this question in section 6.5.

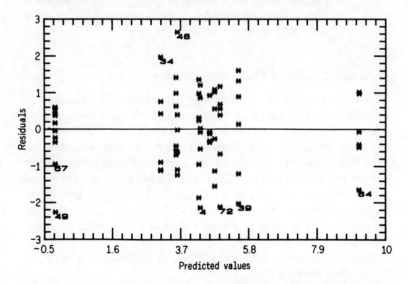

Figure 5.2 *Bacterial clearance data. Residuals and logarithms of the predicted values. Residuals have been standardized by the MAD. Transform both sides with variance model A. Maximum-likelihood estimator. $\mu = -0.05$. Selected cases indicated.*

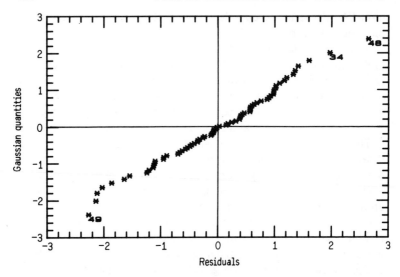

Figure 5.3   *Bacterial clearance data. Normal probability plot of residuals. Residuals have been standaridzed by the MAD. Transform both sides with variance model A. Maximum-likelihood estimator. $\mu = -0.05$. Selected cases indicated.*

Example 5.2   Skeena River sockeye salmon

This is a continuation of Example 4.1 in section 4.5. In section 4.5 these data were analyzed by transforming both sides, and a conflict was found between transforming to symmetry and transforming to heteroscedasticity. The transformation to zero skewness was $\lambda_{sk} = 0.45$, the transformation to homoscedasticity was $\hat{\lambda}_{het} = -0.86$, and the MLE was -0.2. Using the MLE removed most but not all of the heteroscedasticity and replaced slight right-skewness by slight left-skewness. Also, prediction intervals for recruitment were somewhat unsatisfactory; for large values of S (spawners) they were too short.

Consider the model combining TBS with weighting by a power-of-the-median response

$$R_i^{(\lambda)} = f_{RK}^{(\lambda)}(S_i, \beta) + \sigma\mu_i^{\alpha}\varepsilon_i \qquad (5.12)$$

where $\mu_i = f_{RK}(S_i, \hat{\beta}_p)$ with $\hat{\beta}_p$ a preliminary estimate. Because λ and α can vary independently, λ can be chosen to transform to symmetry while α corrects the heteroscedasticity. A slight variant of (5.12) is to replace μ_i^α by S_i^θ. Since μ_i is a monotonic increasing function of S_i, this change will not have a great effect.

Both types of weighting can be combined into a single model

$$R_i^{(\lambda)} = f_{RK}^{(\lambda)}(S_i, \beta) + \sigma \mu_i^\alpha S_i^\theta \varepsilon_i \tag{5.13}$$

We will use model (5.13) only with the constraint $\theta = 0$ or the constraint $\alpha = 0$. Even with one of the parameters α and θ fixed, the other's estimator is highly correlated with $\hat{\lambda}$; this is a problem but not insurmountable. Allowing α, λ, and θ to vary simultaneously would be an overparameterization; the information matrix would be near-singular and $\hat{\alpha}$, $\hat{\lambda}$, and $\hat{\theta}$ would be highly correlated with large standard errors.

Interesting special cases of model (5.13) were fit. When power weighting by μ_i, the TBS estimator was used as the preliminary estimate $\hat{\beta}_p$. In Table 5.3 these models are compared. TBS, corresponding to $\alpha = \theta = 0$, has the smallest maximized loglikelihood. The

Table 5.3 *Skeena River sockeye salmon data. TBS and power weighting models*

Model	Estimates (standard errors)			2 × (loglikelihood ratio)
	$\hat{\lambda}$	$\hat{\alpha}$	$\hat{\theta}$	
(1) TBS and weighting by μ_i^α	0.39 (0.34)	0.99 (0.42)	*	0
(2) TBS and weighting by S_i^θ	0.36 (0.34)	*	0.68 (0.26)	0.02
(3) Weighting by μ_i^α	*	1.93 (0.65)	*	1.5
(4) Weighting by S_i^θ	*	*	1.09 (0.20)	1.9
(5) TBS	− 0.2 (0.28)	*	*	3.44

*Parameter fixed at its default value, which is 1 for λ and 0 for α and θ.

largest loglikelihood is achieved by TBS with weighting by a power of μ_i^α, but this model is virtually indistinguishable from TBS and weighting by S_i^θ.

Suppose we fix $\theta = 0$ and wish to test H_0: $\alpha = 0$. Twice the loglikelihood ratio is 3.44 (see Table 5.3). Comparing this to chi-squared values on one degree of freedom, one rejects H_0 at $p = 0.1$ but not at $p = 0.05$. This suggests that TBS alone just barely fits the data compared to TBS plus power weighting. The problems we saw when applying TBS to these data may be real; there seems to be a conflict, albeit slight, between transforming to symmetry and homoscedasticity. However, Table 5.3 shows that power weighting alone, using either μ_i^α or S_i^θ, provides an acceptable fit.

When TBS is applied with power weighting, then $\hat{\lambda}$ is close to $\hat{\lambda}_{sk} = 0.45$, showing that the skewness alone is determining λ.

How do the prediction intervals for recruitment given by weighting alone or weighting and TBS combined compare with the prediction intervals by TBS alone? Table 5.4 gives the 90% prediction limits for low, typical, and high spawner values by the three methods.

The lowest spawner value is an isolated point. It is case 16, four years after the rockslide year. Since the point is isolated, it is impossible to tell which of the intervals at that spawner value is most realistic.

The prediction intervals at $S = 511$ by TBS or TBS and weighting seem realistic when compared to Figure 4.1. The upper limit by weighting alone is somewhat low; the symmetric intervals generated by this method do not fit the data as well as the asymmetric intervals generated by TBS.

Table 5.4   *Skeena River sockeye salmon. 90% prediction intervals for recruitment. Ricker model. The three spawner values are the lowest observed value, a typical value, and the highest observed value*

Method	Spawners (thousands of fish)		
	87	511	1066
TBS	185, 524	625, 2473	749, 3142
TBS and weight by S^α	222, 413	605, 2182	484, 3566
Weight by S^α	194, 415	526, 2050	52, 3450

At $S = 1066$, the interval by TBS and weighting conforms to the data much better than the other two intervals. The lower limit by weighting is far too low; again the symmetry of the interval is to blame. The lower limit by TBS is too high; this problem has been discussed in section 4.5.

For most purposes, either TBS or power weighting would be adequate, but power weighting might be preferable. If more data were available, then we might want to combine TBS with power weighting; even with only 27 data points this model is quite reasonable and gives the most realistic prediction limits.

CHAPTER 6

Influence and robustness

6.1 Introduction

The classical theory of parametric statistics tells how to test and estimate efficiently when the data are generated exactly by the parametric model. In data analysis, however, there is always the danger that minor problems with the data or the model will destroy the good properties of the classical estimators. Even seemingly slight deficiencies in the data or small modeling errors can have disastrous effects. As many authors have pointed out, a single outlier in a large data set can overwhelm the normal-theory maximum-likelihood estimator. Also, when the errors in a regression model are close to normally distributed but with heavier tails, then the least-squares estimator can be substantially less efficient than certain alternative estimators. The latter include robust estimators and the maximum-likelihood estimator using the actual distribution of the errors when this distribution is known.

Data inadequacies are often caused by gross measurement or recording errors. As Hampel *et al.* (1986, sec. 1.2c) document, routine data contain about 1–10% gross errors, and even the highest-quality data cannot be guaranteed free of gross errors.

In regression analysis, modeling errors can include misspecification of the mean or median function, an incorrect variance function, and the incorrect assumption of normal errors. In recent years, a variety of statistical tools have become available to handle data or modeling inadequacies. The most important tools are robust statistical methods and influence diagnostics.

Robust methods are designed to work well when the probability mechanism generating the data is only close to the assumed parametric model. When the model can provide a satisfactory fit to the bulk of the data, then a good robust estimator will find this fit and will locate the few cases fitting poorly. In contrast, the classical maximum-likelihood estimator will try to fit the model to *all* the data even if this cannot be done well.

Robust estimators can handle both data and model inadequacies. They will downweight and, in some cases completely reject, grossly erroneous data. In many situations, a simple model will adequately fit all but a few unusual observations. For example, a model that is linear in a covariate x and ignores another covariate z will be adequate for a limited range of x and nearly constant value of z. If new data with unusual values of x or z are added, then this model may no longer suffice. If the new data are few, then the best strategy may be retaining the simple model, but using a robust estimator to downweight or reject the unusual cases.

Although robust methods are designed to minimize the effects of problem data or modeling inadequacies, they are not specifically designed to analyze these problems, only to mitigate them. Diagnostics, on the other hand, are designed to tell us *why* certain cases are unusual and have a large influence on the analysis. They also can tell us *how* a model may be inadequate. Outliers or modeling errors, when properly analyzed, may lead to new understanding and may suggest fruitful areas for further research.

With some exceptions, researchers in the field of robustness have ignored the diagnostics literature and vice versa. This is unfortunate and has caused a lack of information on using these tools together. Neither diagnostics nor robust methods alone are as useful as the intelligent combination of both. Robust estimators and tests will often provide diagnostics as byproducts and these can be quite useful, but they are not substitutes for specifically designed diagnostics. It is often as important to analyze the outliers as it is to fit the model to the remaining data. Using diagnostics we can learn much about the outliers that will not be apparent from the robust fit. Robust methods are not a substitute for diagnostics.

Conversely, diagnostics cannot replace good robust methods. Although deleting outliers detected by diagnostics can greatly improve the normal-theory maximum-likelihood and other classical estimators, data deletions cannot achieve the same high efficiency as good robust estimators (Hampel, 1985).

Extreme care is needed when deleting or downweighting data. In linear regression, practitioners often try robustifying the least-squares estimate in the following manner. The residuals from a least-squares fit are examined, outliers are removed, and the least-squares estimate is recomputed from the remaining cases. Because the criteria for rejection are subjective and often vague, this procedure is difficult to study theoretically. Ruppert and Carroll (1980) studied a variant where the outliers are rejected by a fixed rule. Residuals from a preliminary estimate, which could be least squares, are computed, and the cases corresponding to fixed percentages of the highest and lowest residuals are trimmed. The estimate is an analog to the trimmed mean of a univariate sample. Surprisingly, the trimmed least squares estimator is neither efficient nor robust if the preliminary estimate is least squares. The point is that seemingly sensible, but *ad hoc*, methods for coping with bad data may fail.

Moreover, the outliers will include valid data as well as gross errors. For this reason the sampling distributions of estimators and test statistics are changed, perhaps substantially, by deletions of data. The nature of these changes has not been studied in detail. The usual methods for calculating standard errors and p-values should not be trusted if data have been deleted. In contrast, the large-sample distributions of robust estimators and tests are well known, having been amply discussed in the literature.

There is a substantial literature on diagnostics and robustness. Books on diagnostics include Belsley *et al.* (1980), Cook and Weisberg (1982), and Atkinson (1985). Robustness is discussed in books by Huber (1981), Hampel *et al.* (1986), and Hoaglin *et al.* (1983, 1985). The last contains a very readable introduction to the subject. Even these books only treat a small portion of the large and rapidly growing robustness and diagnostics literature.

In a single chapter it would be impossible to discuss diagnostics and robustness in a comprehensive manner. Instead, we describe some simple methods for generating diagnostics and for modifying estimators to make them robust (i.e., to 'robustify' them). These methods are very flexible. They can be applied to any of the estimators of variance functions that are discussed in Chapters 2 and 3 and to transformation models with or without a variance function.

Sections 6.2 and 6.3 introduce these general diagnostic and robustification methods. In sections 6.4 and 6.5 these methods are specialized to weighting by variance functions, transformations, and the combination of weighting and transformation.

6.2 General methods for diagnostics

The earlier diagnostic methods measured the effects upon the statistical analysis of deleting individual cases or groups of cases. For example, Cook's (1977) distance and the statistic DFFITS (Belsley et al., 1980) measure the effect of case deletion upon the least-squares fitted values from a linear regression model. Belsley et al. (1980) propose many other statistics for analyzing the effects of case deletion in linear regression. For example, their statistic DFBETAS is a rescaling of the change in the least-squares estimate upon case deletion.

Some of these single-case-deletion diagnostics have been generalized to group deletions. In principle, such generalizations are not difficult. However, group deletion can be unwieldy in practice because even small data sets have an extremely large number of subsets for possible deletion, and emphasis has been on single-case-deletion diagnostics. There are ways, however, to identify a small collection of subsets that should include any influential subset. For example, Atkinson (1986b) suggests using a robust fit to identify a group of possible outliers, and then applying a group-deletion diagnostic to this group.

Case-deletion diagnostics are practicable for linear models because simple computational formulas are available, so it is not necessary actually to recompute the estimator after deleting each case. These formulas employ algebraic identities that apply to linear estimators. To use them on a nonlinear estimator, one must approximate that estimator by a linear one.

Cook (1986) emphasizes that case deletion is only one of a variety of ways to perturb the data, the model, or the estimator. Other types of perturbations may give us different diagnostic information and may be easier to compute as well. In particular, Cook advocates the assessment of 'local influence', that is the effects of infinitesimally small perturbations. He studies the normal linear model and, as examples, analyzes the effects caused by changes in the assumed variances of the observations and by changes in the values of the explanatory variables.

In an ordinary regression model, a change in the variance of an observation is equivalent to a change in the reciprocal of the case weight. However, when the model includes a nonconstant variance function or a transformation parameter, a change in case weight cannot be interpreted simply as a change in the variance of that case. Nonetheless, minor perturbations of case weights are easy to analyze and give useful information about influence.

Section 7.1 outlines the theory of M-estimators for independent observations. M-estimators are very general and include all estimators used in this book. We now turn our attention to the effects of case-weight perturbations on M-estimators.

Consider a p-dimensional parameter and the M-estimator $\hat{\theta}$ satisfying the estimating equation

$$\sum_{i=1}^{N} \Psi_i(y_i, \hat{\theta}) = 0 \tag{6.1}$$

Here $\Psi_i(y, \theta)$ is a p-dimensional 'score function'. Equation (6.1) becomes the 'likelihood equation' when Ψ_i is the Fisher score function, i.e., the gradient of the loglikelihood of y_i.

If we change the weight of the jth case from 1 to $(1 + \varepsilon)$, then by a Taylor series of $\hat{\theta}_\varepsilon$ about $\hat{\theta}$, the new value $\hat{\theta}_\varepsilon$ of the estimator satisfies

$$0 = \sum_{i=1}^{N} \Psi_i(y_i, \hat{\theta}_\varepsilon) + \varepsilon \Psi_j(y_j, \hat{\theta}_\varepsilon)$$

$$\simeq \left(\sum_{i=1}^{N} \dot{\Psi}_i(y_i, \hat{\theta}) + \varepsilon \dot{\Psi}_j(y_j, \hat{\theta}) \right)(\hat{\theta}_\varepsilon - \hat{\theta})$$

$$+ \sum_{i=1}^{N} \Psi_i(y_i, \hat{\theta}) + \varepsilon \Psi_j(y_j, \hat{\theta}) \tag{6.2}$$

Here $\dot{\Psi}_i$ is the derivative of Ψ_i with respect to θ. Using (6.1) and (6.2) we have the following measure of the influence of the jth case

$$(\hat{\theta}_\varepsilon - \hat{\theta})/\varepsilon \simeq -\left(\sum_{i=1}^{N} \dot{\Psi}_i(y_i, \hat{\theta}) + \varepsilon \dot{\Psi}_j(y_j, \hat{\theta}) \right)^{-1} \Psi_j(y_j, \hat{\theta}) \tag{6.3}$$

Because (6.3) measures the relative change in $\hat{\theta}$ as the weight of the jth case is changed, it is roughly the effect that case j has on $\hat{\theta}$.

Since only one observation out of N has been modified, it is better to rescale (6.3) by a multiplicative factor of N; this makes the change from $\hat{\theta}$ to $\hat{\theta}_\varepsilon$ almost independent of N, at least for large N.

The approximate equality in (6.3) becomes exact as ε tends to 0, so that

$$\lim_{\varepsilon \to 0} [N(\hat{\theta}_\varepsilon - \hat{\theta})/\varepsilon] = -\hat{B}^{-1} \Psi_j(y_j, \hat{\theta}) \tag{6.4}$$

where

$$\hat{B} = N^{-1} \sum_{i=1}^{N} \dot{\Psi}_i(y_i, \hat{\theta}) \tag{6.5}$$

The limit in (6.4) is similar to Hampel's (1974) notion of the influence function. The influence function is most easily defined when the y are identically distributed and if Ψ_i is independent of i. Then the influence function is the relative change in $\hat{\theta}$ as an infinitesimally small amount of contamination at y is added to a large sample taken from the distribution of the y. A short calculation (Huber 1981, p. 45) shows that the influence function is

$$\text{IF}(y, \theta) = - [E(\dot{\Psi}(y, \theta))]^{-1} \Psi(y, \theta) \tag{6.6}$$

When the y are not identically distributed, then a natural generalization of (6.6) is

$$\text{IF}_i(y_i, \theta) = - B_N^{-1} \Psi_i(y_i, \theta) \tag{6.7}$$

where, as in equation (7.7), B_N is defined by

$$B_N = E\left(N^{-1} \sum_{i=1}^{N} \dot{\Psi}_i(y_i, \theta) \right) \tag{6.8}$$

If we replace B_N by $\hat{B}$ and θ by $\hat{\theta}$ in (6.7), then we obtain the right-hand side of (6.4), which we will call the empirical influence function

$$\text{EIF}_j = - \hat{B}^{-1} \Psi_j(y_j, \hat{\theta}) \tag{6.9}$$

Cook and Weisberg (1982, sec. 3.2) use similar terminology.

We now turn to case deletions. Letting $\varepsilon = -1$ is equivalent to deleting the jth case. Define $\hat{\theta}_{(j)}$ to be the estimate without the jth case and let $\Delta^E \theta_j = (\hat{\theta} - \hat{\theta}_{(j)})$. By (6.3), $\Delta^E \theta_j$ is approximately equal to $\Delta^A \theta_j$ defined by

$$\Delta^A \theta_j = - [\hat{C}_{(j)}]^{-1} \Psi_j(y_j, \hat{\theta}) \tag{6.10}$$

where

$$\hat{C} = \sum_{i=1}^{N} \dot{\Psi}_i(y_i, \hat{\theta}) = N\hat{B}$$

and

$$\hat{C}_{(j)} = \hat{C} - \dot{\Psi}_j(y_j, \hat{\theta})$$

Formula (6.10) requires a matrix inversion for each case. This is acceptable for small N. For large N, a further approximation can be used. Using this approximation only a single matrix inversion is needed, one to compute $\hat{C}^{-1}$. Let $d_j = \dot{\Psi}_j(y_j, \hat{\theta})$. Then

$$\hat{C}_{(j)}^{-1} = [\hat{C}(I - \hat{C}^{-1}d_j)]^{-1} = (I - \hat{C}^{-1}d_j)^{-1}\hat{C}^{-1}$$

For large N, $\hat{C}^{-1}d_j$ will be small and the geometric series

$$(I - \hat{C}^{-1}d_j)^{-1} = \sum_{k=0}^{\infty} (\hat{C}^{-1}d_j)^k$$

will converge. Covergence is guaranteed if the matrix norm of $\hat{C}^{-1}d_j$ is less than 1, but we will not go into such mathematical issues here. Using a truncation of this series we have

$$[\hat{C}_{(j)}]^{-1} \simeq \sum_{k=0}^{n} (\hat{C}^{-1}d_j)^k \hat{C}^{-1} \tag{6.11}$$

Substituting (6.5) with $n = 1$ into (6.10) gives us

$$\Delta^A\theta_j \simeq -[\hat{C}^{-1} + \hat{C}^{-1}d_j\hat{C}^{-1}]\Psi(y_j, \hat{\theta}) \tag{6.12}$$

If instead we truncate at $n = 0$, then we obtain the empirical influence function, (6.9).

As a diagnostic we can use the local influence diagnostic EIF_i, the exact case-deletion diagnostic $\Delta^E\theta_j$, or the approximate case-deletion diagnostic $\Delta^A\theta_j$. It is difficult at this time to make strong recommendations about which to use. They have not been compared in the literature, and since they are relatively new, there has been little practical experience with them.

In linear regression $\Delta^E\theta_j$, which is DFBETA in the notation of Belsley *et al.* (1980), has become traditional – actually it is the scaled version DFBETAS that is now in common use. For this reason, when using nonlinear estimators statisticians may be most comfortable with the approximation $\Delta^A\theta_j$. However, our tentative recommendation is to use local influence diagnostics because they are easy to compute. A problem with EIF_i is that it does not control for leverage; a high-leverage case may not be flagged even if it is a severe response outlier. High-leverage points can mask themselves. However, leverage and the masking of one observation by another can both be controlled with a robust estimator. In the next section we introduce a local influence diagnostic calculated at a robust estimator. This diagnostic seems particularly informative and easy to use. If only one diagnostic is to be computed, this one may be the best choice presently available.

One can use $\Delta^A\theta_j$ (or $\Delta^E\theta_j$) directly or one can rescale by the standard errors of the components. Which is best? Largely this is a matter of taste. A change in a parameter measured by $\Delta^A \theta_j$ will probably not be important if the change is small relative to the standard error. Thus a scaled diagnostic, such as DFBETAS, helps us decide when high influence is truly important. On the other hand, when a standard error is quite small, a large relative change may still be of little practical importance. This situation is analogous to testing for a difference between two treatments, where a difference could be statistically significant but of no practical significance.

6.3 Robust estimation

From equations (6.7) and (6.9), one can see that the influence function and empirical influence functions of an M-estimator are unbounded if the function $\Psi_i(y, \theta)$ is unbounded. This means that even a single bad observation can have a large influence on $\hat{\theta}$.

We can define a robust estimator $\tilde{\theta}$ by replacing the estimating equation (6.1) with

$$0 = \sum_{i=1}^{N} w_i(y_i, \tilde{\theta})[\Psi_i(y_i, \tilde{\theta}) - \alpha] \qquad (6.13)$$

where $w_i(y, \theta)$ is a weighting function with values between 0 and 1. In general, downweighting by w_i can introduce bias, and the constant α is needed to correct for this bias. In many common situations, e.g., regression with symmetric errors, symmetry considerations show that weighting by w_i does not create bias, and then the shift α will be 0.

Define $\xi_i(y, \theta) = w_i(y, \theta)[\Psi_i(y, \theta) - \alpha]$. The influence function $\mathrm{IF}_i(y, \theta)$ is given by (6.7) and (6.8) with Ψ_i replaced by ξ_i.

How should we choose w_i? The weighting by w_i must provide a balance between robustness and efficiency. Presumably Ψ_i has already been chosen to have good efficiency at the ideal parametric model, and to retain this efficiency $w_i(y, \theta)$ should be equal to 1 as much as possible. For robustness we will choose $w_i(y, \theta)$ so that the influence function $\mathrm{IF}_i(y, \theta)$ is bounded as a function of i and y.

The influence function will be bounded if the so-called gross-error sensitivity is finite. When θ is univariate the gross-error sensitivity, defined as

$$\gamma = \sup_i \sup_y \{|\mathrm{IF}_i(y, \theta)|\}$$

measures the worst effect that a contaminating observation can have on $\tilde{\theta}$.

Hampel (1968, lemma 5) shows how to choose the weighting function $w_i(y, \theta)$ in order to minimize the asymptotic variance of $\tilde{\theta}$ subject to a bound b on γ; the lemma can also be found in Hampel et al. (1986, p. 117). Except that the choice of b is still open, Hampel's lemma states precisely how one can achieve an optimal compromise between robustness against deviations from the ideal model and efficiency at the model. We will not discuss this lemma further since it applies only to univariate parameters and therefore has little applicability in regression analysis. Instead we turn directly to multivariate parameters.

In the multidimensional case, the optimal choice of w_i is not so unambiguous. When $\tilde{\theta}$ is multivariate there are several ways to generalize the definition of gross-error sensitivity as well as various measures of the size of the asymptotic covariance matrix. These are discussed in detail by Krasker and Welsch (1982) and Hampel et al. (1986, chap. 4). Here we can only summarize those discussions.

The unstandardized gross-error sensitivity is

$$\gamma_u = \sup_i \sup_y \{ \| IF_i(y, \theta) \| \}$$

where $\| \cdot \|$ is the ordinary Euclidean norm. Because the individual components of IF_i depend on the parameterization, γ_u is not invariant to reparameterization. For example, in linear regression if one replaces a covariate x_j by Kx_j, $K \neq 0$, then the jth component, β_j, of the regression parameter changes to β_j/K. The jth component of IF_i is also divided by K. As $K \to 0$, γ_u becomes dominated by the jth component. Conversely as $K \to \infty$ the jth component of IF_i tends to 0 and γ_u becomes insensitive to the influence for β_j. In summary, γ_u is a suitable measure of sensitivity to outliers only if the parameters have been properly scaled to be comparable.

Instead of attempting to scale the parameters, we will use a measure of gross-error sensitivity that is invariant to reparameterization. The 'information-standardized' and 'self-standardized' gross-error sensitivities have this property (Hampel et al. 1986, sec. 4.2b). The former uses a quadratic form in the Fisher information matrix, which is itself sensitive to outliers. For this reason we will restrict ourselves to the self-standardized gross-error sensitivity.

Krasker and Welsch (1982) introduced the self-standardized gross-error sensitivity and obtained the first estimator bounding this sensitivity. Their work was confined to normal-theory linear regression. We will now generalize their ideas to arbitrary parametric families.

Let $V(\theta)$ be the asymptotic covariance matrix of $N^{1/2}(\tilde{\theta} - \theta)$ when θ is the true parameter. The self-standardized gross-error sensitivity is so-called because it uses $V(\theta)$ to norm the influence function of $\tilde{\theta}$. The following notation will be useful; for any $p \times p$ positive definite matrix M the norm $\|\cdot\|_M$ on $\mathbb{R}^p$ is defined by

$$\|x\|_M = [x^\mathrm{T} M^{-1} x]^{1/2}$$

Then the self-standardized gross-error sensitivity is defined as

$$\gamma_s = \sup_i \sup_y \| \mathrm{IF}_i(y, \theta) \|_{V(\theta)} \qquad (6.14)$$

It is easy to see that γ_s is invariant to reparameterization. Let κ be a one-to-one transformation of the parameter space onto itself. If we transform θ to $\kappa(\theta)$, then $\mathrm{IF}_i(y_i, \theta)$ and $V(\theta)$ are transformed to $\mathrm{IF}_i(y_i, \kappa) = \dot{\kappa}(\theta)\mathrm{IF}_i(y_i, \theta)$ and $V(\kappa) = \dot{\kappa}(\theta)V(\theta)[\dot{\kappa}(\theta)]^\mathrm{T}$, respectively, and γ_s remains unchanged.

It is instructive to consider the case where $V(\theta)$ is diagonal. We know that as the parameters are rescaled, γ_s remains constant but γ_u changes. If each parameter is rescaled through division by the standard deviation of its estimator, then γ_u will equal γ_s.

When $V(\theta)$ is not diagonal, one can use the reparameterization $\theta \to \kappa(\theta) = [V(\theta)^{-1/2}\theta]$. The asymptotic covariance matrix of κ is the identity, and the unstandardized and self-standardized sensitivities of κ are equal. For a further discussion and a somewhat different motivation of the self-standardized sensitivity, see Krasker and Welsch (1982).

By equation 7.9 $V(\theta) = B^{-1}AB^{-\mathrm{T}}$ where B and A are the limits of B_N and A_N given by (6.8) and

$$A_N = E\left(N^{-1} \sum_{i=1}^N \Psi_i(y_i, \theta)\Psi_i^\mathrm{T}(y_i, \theta) \right)$$

respectively. Using (6.7) and replacing $V(\theta)$ in (6.14) by $(B_N^{-1} A_N B_N^{-\mathrm{T}})$ we have a finite-N version of γ_s

$$\gamma_{s,N} = \sup_i \sup_y \| \xi_i(y, \theta) \|_{A_N} \qquad (6.15)$$

It is known (Krasker and Welsch, 1982) that $\gamma_{s,N}$ must be at least $\sqrt{p}$. It has been our experience that bounding $\gamma_{s,N}$ by $a\sqrt{p}$ where a is between 1.2 and 1.6 achieves the dual goals of high efficiency at the ideal model and insensitivity to outliers. Since $\gamma_{s,N}$ depends upon unknown parameters, we bound instead the estimate

$$\hat{\gamma}_s = \sup_i \sup_y \| \xi_i(y_i, \tilde{\theta}) \|_{\hat{A}}$$

$$= \sup_i \sup_y \| w_i(y_i, \tilde{\theta})[\Psi_i(y_i, \tilde{\theta}) - \alpha] \|_{\hat{A}} \qquad (6.16)$$

where

$$\hat{A} = N^{-1} \sum_{i=1}^{N} \xi_i(y_i, \tilde{\theta})\xi_i^{\mathrm{T}}(y_i, \tilde{\theta})$$

$$= N^{-1} \sum_{i=1}^{N} w_i^2(y_i, \tilde{\theta})[\Psi_i(y_i, \tilde{\theta}) - \alpha][\Psi_i(y_i, \tilde{\theta}) - \alpha]^{\mathrm{T}}$$

$$(6.17)$$

It follows from (6.16) that w_i will be as large as possible, subject to $\hat{\gamma}_s \leqslant a\sqrt{p}$ if

$$w_i(y_i, \tilde{\theta}) = \min \{1, (a\sqrt{p})/\| \Psi_i(y_i, \tilde{\theta}) - \alpha \|_{\hat{A}} \qquad (6.18)$$

There is another form for the weights that will be of interest soon. Let ψ be Huber's (1964) 'psi function' defined by

$$\psi(x) = \begin{cases} x & \text{if } |x| \leqslant a\sqrt{p} \\ (a\sqrt{p})\mathrm{sgn}(x) & \text{otherwise} \end{cases}$$

Then

$$w_i(y_i, \tilde{\theta}) = \psi(\| \Psi_i(y_i, \tilde{\theta}) \|_{\hat{A}})/\| \Psi_i(y_i, \tilde{\theta}) \|_{\hat{A}} \qquad (6.19)$$

The estimates $\tilde{\theta}$, $\hat{A}$, and the w_i must be found simultaneously by solving (6.13), (6.17), and (6.18). To solve these equations, we have developed an iterative algorithm, which will be described shortly.

By construction, $\tilde{\theta}$ achieves the robustness requirement that $\hat{\gamma} \leqslant a\sqrt{p}$, and for this reason will be called a 'bounded-influence' estimator.

In what sense is $\tilde{\theta}$ optimal at the true model? If Ψ_i is the Fisher score function, so that the unweighted estimator is the maximum-likelihood estimator, then we can expect $\tilde{\theta}$ to be reasonably efficient. However, the issue of optimality is complex. One case has been discussed in detail, normal-theory linear regression. In this case $\tilde{\theta}$ is the so-called Krasker and Welsch (1982) estimator. Krasker and Welsch conjectured that, within a certain class of estimators satisfying a fixed bound on γ_s and some other conditions, $\tilde{\theta}$ was strongly optimal. Strong optimality means that at the assumed model $\tilde{\theta}$ minimizes the asymptotic variance of every linear combination $\zeta^T\tilde{\theta}, \zeta$ in $\mathbb{R}^p$, or equivalently that $\tilde{\theta}$ minimizes the asymptotic covariance matrix in the sense of positive definiteness.

Ruppert (1985) constructed a counterexample showing that the Krasker–Welsch conjecture was false. The point of the counter-example is that we can improve over $\tilde{\theta}$ for certain components of θ, if we are willing to sacrifice efficiency for other components. Such a trade-off might be useful if only certain parameters are of primary interest and the others are nuisance parameters. However, in many situations $\tilde{\theta}$ is almost as efficient at the ideal model as the maximum-likelihood estimator, so little improvement over $\tilde{\theta}$ is possible. See Ruppert (1985) for a fuller discussion and examples.

Hampel *et al.* (1986, sec. 4.3, theorem 2) and Stefanski *et al.* (1986, theorem 1) show that $\tilde{\theta}$ achieves a certain type of admissibility. Among all estimators with the same gross-error sensitivity, there exists no estimator that has a strictly smaller asymptotic covariance matrix. Admissibility means that one can improve over $\tilde{\theta}$ for certain components of θ only if one sacrifices efficiency for other components. Thus, admissibility is the converse of Ruppert's (1985) counter-example.

Large-sample tests and confidence regions for the parameters can be constructed using an estimate $\hat{V}$ of the asymptotic covariance matrix of $N^{1/2}(\tilde{\theta} - \theta)$. From section 7.1, a suitable estimate is $\hat{V} = \hat{B}^{-1}\hat{A}\hat{B}^{-T}$, where $\hat{A}$ is the solution to (6.17) and

$$\hat{B} = N^{-1} \sum_{i=1}^{N} \dot{\xi}_i(y_i, \tilde{\theta}) \qquad (6.20)$$

which is the same estimate of B given by equation (7.12). When the derivative $\dot{\xi}$ is difficult to calculate analytically, one can either use a numerical derivative or replace $\hat{B}$ by B^* given by (7.13). We prefer using $\hat{B}$ with numerical derivatives since B^* is consistent for B only at the ideal model. One should be cautious when using these tests. They suffer from the inaccuracy of all Wald-type tests and confidence regions.

Algorithm for the bounded-influence estimator We have had good success with the following algorithm for computing $\tilde{\theta}$, $\hat{A}$, and the w_i. We have only used it in situations where the shift α was 0, or at least α seemed sufficiently small that it could be taken as 0. This seems to be true for all estimators covered in this book. Therefore, the algorithm does not estimate α. The algorithm is as follows.

Step 1   Fix $a > 1$. Let C be the total number of iterations that will be used. Set $c = 1$. Let $\tilde{\theta}_p$ be a preliminary estimate, possibly the MLE. Another possibility is to let $\tilde{\theta}_p$ be an estimator with a high breakdown point; see the discussion of breakdown below. Set $w_i = 1$ for all i.

Step 2   Define

$$\hat{A} = N^{-1} \sum_{i=1}^{N} w_i^2 \Psi_i(y_i, \tilde{\theta}_p) \Psi_i^{\mathrm{T}}(y_i, \tilde{\theta}_p)$$

Step 3   Using (6.18) update the weights

$$w_i = \min \{ 1, (a\sqrt{p}) / \| \Psi_i(y_i, \tilde{\theta}_p) \|_{\hat{A}} \}$$

Step 4   Using these fixed weights solve the following equation for $\tilde{\theta}$

$$0 = \sum_{i=1}^{N} w_i \Psi_i(y_i, \tilde{\theta})$$

In most applications that concern us, Ψ_i is the Fisher score function, and step 4 can be solved using a weighted scoring algorithm. Of course, a Newton or quasi-Newton algorithm would also be suitable.

Step 5   If $c < C$, set $\tilde{\theta}_p = \tilde{\theta}$, $c = c + 1$, and return to step 2. Otherwise, stop.
End

The convergence of the algorithm can be slow if there are severe outliers and $\tilde{\theta}_p$ is the maximum-likelihood estimator. It has been our experience that about 10 iterations are needed before the weights given to severe outliers stop decreasing. In such cases, the number of iterations could be decreased if a more robust preliminary estimator were available. In examples where there are no severe outliers, the algorithm typically stabilizes after two or three iterations.

One danger is that the slow convergence in the presence of severe outliers can give the appearance that the estimator has stabilized after only two or three iterations, when in fact it is still changing as the weights of the outliers slowly decrease.

We implemented this algorithm on the matrix language GAUSS. When Ψ was the gradient of the loglikelihood we used numerical derivatives for convenience. Numerical second derivatives were used to calculate $\hat{B}$.

6.3.1 Redescending estimators

Extreme outliers are often gross errors. Hampel *et al.* (1986) discuss the need for robust estimators that completely reject extreme outliers instead of merely bounding their influence. For location parameters extreme outliers can be automatically rejected by using a redescending 'psi function'.

We will introduce an analog to a redescending location estimator. Let $\psi(x)$ be an odd function with $\psi(x) \geqslant 0$ for $x \geqslant 0$. We saw that the weights defined by (6.18) are also given by (6.19) when ψ is Huber's psi function. To be more flexible in the choice of weights, we can instead define the weights by (6.19) using a general psi function. If for some $R > 0$, $\psi(x) = 0$ for all $|x| \geqslant R$, then extreme outliers are completely rejected.

In the following examples we used Hampel's three-part redescending psi function

$$\psi(x) = \begin{cases} x & 0 \leqslant x \leqslant b_1 \\ b_1 & b_1 \leqslant x \leqslant b_2 \\ b_1(b_3 - x)/(b_3 - b_2) & b_2 \leqslant x \leqslant b_3 \\ 0 & b_3 \leqslant x \end{cases} \tag{6.21}$$

where $(b_1, b_2, b_3) = (\sqrt{p})(a_1, a_2, a_3)$. Typically a values are $1.3 \leqslant a_1 \leqslant 1.75, 3.0 \leqslant a_2 \leqslant 4.5$, and $6 \leqslant a_3 \leqslant 9$. It is worth experimenting with the constant a of the bounded-influence estimator and the constant a_1, a_2, and a_3 of the redescending estimators. If the estimates or their standard errors from $\hat{V}$ are sensitive to the choice of these constants, then the data analyst should suspect problems with the data or the model.

6.3.2 Breakdown

We have not discussed the breakdown point of the bounded-influence estimator. Hampel (1971) introduced the idea of breakdown. Roughly speaking, the breakdown point of an estimator is the largest proportion of contamination that the estimator can tolerate before its value can be determined by the contamination. A robust estimator should have a high breakdown point as well as a bounded-influence function.

The influence function and the breakdown point can be viewed as complementary. The influence function only measures the influence of infinitesimal contamination by outliers, but at least for M-estimators of location it approximates well the effects of contamination up to about one-fourth the breakdown point and is useful to about one-half the breakdown point (Hampel et $al.$, 1986). Ideally, a robust estimator should have a bounded-influence function and as high a breakdown point as possible.

We have not investigated the breakdown point of the bounded-influence estimator. However, since the bounded-influence estimator uses an estimated second-moment matrix, $\hat{A}$, its breakdown point cannot be greater than p^{-1}. This follows from Maronna's (1976) work on M-estimation of dispersion matrices. The low breakdown point of bounded-influence estimators should not rule out their use. Typically these estimators can handle data exhibiting multiple outliers and masking; see the examples in Carroll and Ruppert (1985).

Recently, several estimators for linear regression have been proposed that have breakdown points near 50% (see Rousseeuw, 1984; Rousseeuw and Yohai, 1984). These estimators have low efficiency at the normal model. Yohai (1987) has proved that by using them as starting values, one can define highly efficient M-estimators with 50% breakdown points. Yohai's estimators do not have a bounded-influence function. Moreover, the extension of his work to nonlinear regression may involve considerable computational difficulties.

Clearly this is an area needing more research. Once computationally feasible, high-breakdown nonlinear regression estimators are found, it should be possible to adapt them to regression with transformation and weighting parameters.

6.3.3 Diagnostics for the robust estimator

Most diagnostics such as Cook's distance or local influence (Cook, 1986) analyze the effects of data or model perturbations on the classical, i.e., nonrobust, estimator. A persistent problem with these diagnostics is the masking of influential observations by other, even more influential, observations. Masking can make data analysis tedious. The most influential points must be identified and deleted, and then the diagnostics must be repeated without them. It is not clear when to stop this process.

Hampel *et al.* (1986, p. 13) suggest in passing that masking could be mitigated by evaluating the diagnostics at a robust fit. Indeed the weights from a bounded-influence estimator can be used to indicate masking; see Carroll and Ruppert (1985, 1987) for examples. Although these weights can help locate masked points, the weights tell us little about the nature of their influence.

Although we could examine the influence function $IF_i(y_i, \tilde{\theta})$ of a robust fit, since the influence function has been constructed to be bounded it will *not* highlight influential points.

We will now describe a diagnostic that is evaluated at a robust fit. Theoretical considerations and our limited practical experience with this diagnostic indicates that it will flag influential points and be relatively immune to the masking. Of course, no diagnostic can be more resistant to masking than the estimator at which it is evaluated.

Recall that the robust estimator we have introduced solves the equation

$$\sum_{i=1}^{N} w_i \Psi_i(y_i, \tilde{\theta}) = 0$$

where the weights w_i depend on the data. If we reweighted an influential point y_i by multiplying $[w_i \Psi_i(y_i, \tilde{\theta})]$ by $(1 + \varepsilon)$, this will have little effect on $\tilde{\theta}$, since w_i would automatically readjust. This reiterates the point we just made; the influence of y_i on the robust estimator is bounded.

Instead we will *fix* the w_i and replace w_j by $(w_j + \varepsilon)$. Then the estimate, say θ_ε^*, solves

$$\sum_{i=1}^{N} \omega_i \psi_i(y_i, \theta_\varepsilon^*) + \varepsilon \psi_j(y_j, \theta_\varepsilon^*) = 0 \tag{6.22}$$

By the same calculation that led to (6.4), we see that

$$\lim_{\varepsilon \to 0} [N(\theta_\varepsilon^* - \tilde{\theta})/\varepsilon] = -\left(N^{-1} \sum_{i=1}^{N} w_i \dot{\Psi}_i(y_i, \tilde{\theta}) \right)^{-1} \psi_j(y_j, \tilde{\theta}) \tag{6.23}$$

This agrees with (6.4) when $w_i \equiv 1$. We will call the right-hand side of (6.23) the local influence diagnostic, LI_j.

It is informative to compare (6.23) with (6.4) and with the influence function of the robust estimate. In (6.4) the influence of y_j can be masked in two ways: (i) Ψ_j is evaluated at a nonrobust estimator, and (ii) $\dot{\Psi}_i$ is *not* weighted by w_i. A large value of $\dot{\Psi}_k$, for some k, will inflate the value of $\hat{B}$. This can cause $\hat{B}^{-1}$ appearing in (6.4) to have some small eigenvalues and can mask an observation y_j such that $\Psi_j(y_j, \tilde{\theta})$ is close to one of the corresponding eigenvectors.

The influence function of the bounded-influence estimator $\tilde{\theta}$ is equal to LI_j times w_j. Therefore, when evaluated at influential points, LI_j is much larger than the influence function of $\tilde{\theta}$. Consequently, LI_j will highlight the influential points even though the influence function of $\tilde{\theta}$ may not. It is interesting to note that although LI_j is an influence function of θ^* satisfying (6.22), this estimator coincides with the BI estimator only at the original data set, not at perturbations of these data.

The idea of evaluating influence diagnostics at a robust estimator is related to an approach of Atkinson (1986b), who deals with the linear model only. Atkinson suggests using Rousseeuw's (1984) least-median-of-squares estimator to detect a group of outliers, applying least squares to the 'good' data, and then evaluating influence diagnostics at this 'trimmed' least-squares estimate. This technique should be a useful data-analytic tool, but a note of caution is in order. Trimmed least-squares estimators where the trimming is based on a preliminary estimator, even a robust one, can have peculiar distributional properties and may be neither efficient nor robust (see Ruppert and Carroll, 1980; Welsh, 1987).

6.4   Weighted regression

Recall the heteroscedastic regression model

$$y_i = f(x_i, \beta) + \sigma g(\mu_i(\beta), z_i, \theta)\varepsilon_i \qquad (6.24)$$

where

$$\mu_i(\beta) = f(x_i, \beta)$$

In the previous section, we discussed general techniques for generating influence diagnostics and bounding the influence function of an estimator. Now we apply these techniques to two important estimation methods for model (6.24), maximum likelihood and pseudo-likelihood (see Chapter 3). Recall that the difference is that maximum-likelihood estimates β and θ simultaneously, while pseudo-likelihood estimates β with θ fixed and vice versa.

6.4.1 Simultaneous estimation of β and θ

Simultaneous estimation of β and θ is simpler to describe than separate estimation, though the latter can be easier to implement. Let $l_i(y_i, \beta, \theta)$ be the log of the conditional density of y_i given x_i and z_i, i.e.

$$l_i(y_i, \beta, \theta) = -\log[2\pi\sigma^2(\beta, \theta)]/2 - r_i^2/[2\sigma^2(\beta, \theta)]$$
$$-\log[g(\mu_i(\beta), z_i, \theta)] \qquad (6.25)$$

where

$$r_i = [y_i - f(x_i, \beta)]/g(\mu_i(\beta), z_i, \theta) \qquad (6.26)$$

and $\sigma^2(\beta, \theta)$ is the weighted maximum-likelihood estimator of σ^2, that is, the weighted average of $r_1^2, \ldots, r_N^2$ with weights $w_1, \ldots, w_N$. These are the weights defined in step 3 of the algorithm for the bounded-influence estimator (see section 6.3).

The purpose of introducing $\sigma(\beta, \theta)$ is to eliminate the nuisance parameter σ. This is done for two reasons, computational stability and increased sensitivity to influence on $\hat{\beta}$ and $\hat{\theta}$. Because σ depends heavily on θ, iterative algorithms that allow σ and θ to vary independently are unstable if the starting values of σ and θ are not chosen with extreme care. It has been our experience that treating σ as a function of θ and β reduces the need for excellent starting values. When σ is eliminated as a free parameter, influence on $\hat{\sigma}$ is ignored and one can concentrate on influence for important parameters. There is no danger that an observation will be downweighted solely for its influence on $\hat{\sigma}$.

The maximum-likelihood estimator simultaneously solves

$$\sum_{i=1}^{N} l_{\beta,i}(y_i, \hat{\beta}, \hat{\theta}) = 0$$

and

$$\sum_{i=1}^{N} l_{\theta,i}(y_i, \hat{\beta}, \hat{\theta}) = 0$$

where $l_{\beta,i}$ and $l_{\theta,i}$ are the partial derivatives of l_i with respect to β and θ. The theory of sections 6.2 and 6.3 can now be applied to

$$\Psi_i(y, \tau) = \begin{bmatrix} l_{\beta,i}(y_i, \beta, \theta) \\ l_{\theta,i}(y_i, \beta, \theta) \end{bmatrix} \qquad (6.27)$$

where

$$\tau = \begin{bmatrix} \beta \\ \theta \end{bmatrix}$$

Example 6.1   Skeena River sockeye salmon

This data set was discussed in sections 4.5 and 5.3. In the latter section, power weighting with and without transformation was considered. Here we will consider only weighting by S_i^α. This model corresponds to (6.24) with $z_i = S_i$, $\theta = \alpha$, and

$$g(\mu_i(\beta), S_i, \theta) = g(S_i, \alpha) = S_i^\alpha \qquad (6.28)$$

As in Chapter 4, the median recruitment is given by the Ricker model, equation (4.35).

In the previous analyses, the year 1951 (case 12 where a rockslide

Table 6.1   *Skeena River sockeye salmon data. Estimators with the variance proportional to a power of S, the size of the spawning population. The standard errors are in parentheses*

Estimator	$\hat{\beta}_1$	$\hat{\beta}_2$	$\hat{\alpha}$
	Parameter		
β and α estimated simultaneously			
MLE (with case 12)	3.27	5.80	1.61
	(0.54)	(2.84)	(0.35)
(without case 12)	3.74	7.72	2.17
	(0.54)	(2.74)	(0.40)
Bounded-influence (one-step)	3.43	6.67	1.65
	(0.62)	(2.90)	(0.54)
(fully iterated)	3.70	8.08	1.87
	(0.57)	(2.72)	(0.48)
Redescending (one step)	3.89	9.11	1.97
	(0.54)	(2.49)	(0.45)
(fully iterated)	3.94	9.30	1.97
	(0.47)	(2.31)	(0.41)
β and α estimated separately			
Redescending (one-step)			
first iteration	3.72	7.87	1.98
	(0.46)	(2.38)	(0.32)
second iteration	3.69	7.73	1.96
	(0.54)	(2.75)	(0.33)

interfered with spawning) was deleted. Now we will include it to demonstrate the diagnostics and robust estimators.

In Table 6.1 six estimators are given, the maximum-likelihood estimator with and without case 12, the one-step and fully iterated bounded-influence estimators, and the one-step and fully iterated redescending estimators. The redescending estimators used the bounded-influence estimator as a preliminary estimate. The 'tuning constants' were $a = 1.5$ for the bounded-influence estimator and $(a_1, a_2, a_3) = (1.5, 3.5, 8)$ for the redescender. The table also gives the standard errors of the estimators.

Clearly case 12 is the most influential point; it is completely rejected by the redescending estimator. This is somewhat surprising. Although we know that case 12 is an anomalous year, it does not appear that outlying in Figure 4.1. This reiterates a point made in Chapter 2. In the presence of heteroscedasticity, outliers can be difficult to detect by ordinary scatterplots.

In Table 6.2 the influence diagnostics are given for cases 4, 5, 12, 16, and 18, the only cases that are downweighted by the robust estimators.

Looking at the approximate case-deletion diagnostics, Δ_i^A, we see that two points have a large influence on $\hat{\alpha}$. Deleting case 12 increases $\hat{\alpha}$, which indicates greater heteroscedasticity when case 12 is removed. This is in agreement with Figure 4.1. In that figure we see that the recruitment variability increases as a function of S. Case 12 is an outlier, and its large residual occurs with a low value of S. Retaining case 12 increases the recruitment variability at the low end of the range of S, but leaves the variability unchanged for other values of S. Therefore, the recruitment variability as a function of S increases more slowly when case 12 is retained.

Deleting case 5 decreases $\hat{\alpha}$, indicating less heteroscedasticity. Because case 5 is an outlier corresponding to a large value of S, this is also in agreement with Figure 4.1.

It is interesting that the approximate case-deletion diagnostics do not show case 4 as influential, especially for $\hat{\alpha}$. The trouble is not with the accuracy of the approximation; the approximate and exact case-deletion diagnostics are similar. The problem is that case 4 is masked by case 12. The local influence diagnostics do show case 4 as influential. Cases 4 and 12 are similar, each having a negative residual and a low value of S. However, case 12 has a much larger absolute residual and masks case 4. The redescending estimate completely

Table 6.2 Skeena River sockeye salmon data. Influence diagnostics with the variance proportional to a power of S. Δ_i^A is the approximate case-deletion diagnostic given by equation (6.12). LI_i is the local influence diagnostic for the robust estimators, see equation (6.12). W_i is the case weight. BI est. is the bounded-influence estimator. Redes. est. is the redescending estimator

Diagnostic	Parameter	Case number				
		4	5	12	16	18
β and α estimated simultaneously						
Δ_i^A	β_1	−0.17	−0.20	−0.33	0.17	0.20
	β_2	−0.65	−1.58	−1.30	0.74	0.59
	α	0	0.24	−0.36	0.02	−0.05
LI_i (at the	β_1	−0.26	−0.12	−0.72	0.21	0.15
fully iterated	β_2	−0.92	−1.22	−2.69	0.89	0.22
BI est.)	α	−0.19	0.38	−1.32	0.15	−0.05
W_i (at the BI est.)		0.79	0.37	0.16	1.00	0.74
LI_i (at the	β_1	−0.35	−0.16	−1.05	0.26	0.14
fully iterated	β_2	−1.22	−1.41	−4.24	1.10	0.02
redes. est.)	α	−0.39	0.47	−2.63	0.19	−0.14
W_i (at the redes. est.)		0.44	0.004	0	0.76	0.64
β and α estimated separately						
LI_i (at the	β_1	−0.23	−0.13	0.57	0.27	0.16
fully iterated	β_2	0.86	1.21	−2.40	1.20	0.26
BI est.)	α	−0.19	0.32	−1.47	0.11	−0.11
LI_i (at the	β_1	−0.30	−0.17	−0.74	0.25	0.17
fully iterated)	β_2	−1.07	−1.42	−2.99	1.09	0.24

rejects case 12, so the influence of case 4 is *not* masked when local influence is evaluated at the redescender.

We now turn to influence for $\hat{\beta}_1$ and $\hat{\beta}_2$. Deleting cases 4, 5, or 12 increases both $\hat{\beta}_1$ and $\hat{\beta}_2$, while deleting cases 16 or 18 has the opposite effect of decreasing both estimates. To understand the ultimate effects on the estimated median response, we should notice that β_1 is the slope of the Ricker function at $S = 0$ and $S = 1/\beta_2$ is the value that maximizes the Ricker function; the maximum is (β_1/β_2) $\exp(-1)$. Therefore, increasing both $\hat{\beta}_1$ and $\hat{\beta}_2$ causes the estimated median to rise more quickly as a function of S and to have its maximum recruitment at a smaller value of S. Decreasing both $\hat{\beta}_1$ and $\hat{\beta}_2$ has the opposite effects. When both $\hat{\beta}$ values increase or both decrease, the maximum recruitment depends on whether or not the relative increase of $\hat{\beta}_1$ is greater than that of $\hat{\beta}_2$.

From Table 6.1 we see that the exact changes in $\hat{\beta}_1$, $\hat{\beta}_2$, and $\hat{\alpha}$ when deleting case 12 are -0.47, -1.92, and -0.56, respectively. Using Δ_i^A, these changes are approximated by -0.33, -1.30, and -0.36, respectively. This level of accuracy seems acceptable, since Δ_i^A is only intended as an influence diagnostic and knowing the exact changes in the parameter estimates is not essential.

6.4.2 Estimating β and θ separately

In Chapter 3 the pseudo-likelihood method of estimation was described. The idea is to estimate β first, possibly by unweighted least squares, and then to estimate θ by maximum likelihood, treating β as known and fixed. Then β can be re-estimated by weighted least squares, treating θ as known and fixed. This process can be repeated a fixed number of times or possibly until convergence.

At each stage of the pseudo-likelihood method, we can use a robust estimator instead of maximum likelihood. Influence diagnostics can be computed at each stage as well. If the pseudo-likelihood or maximum-likelihood estimator is available, it can be used at the first stage as a preliminary estimate of β; although it is nonrobust it should be more efficient than unweighted least squares.

In Giltinan *et al.* (1986) this robust modification of pseudo-likelihood is discussed in detail. Besides the bounded-influence estimator generalizing the Krasker–Welsch estimate, a generalization of the Mallows estimator is discussed. The Mallows estimate separate influence into two parts, the influence of the residual and the influence due to leverage. For the estimation of β, the influence for leverage is the influence due to x, and there is an analogous meaning for leverage when estimating θ.

The early robust estimators of Carroll and Ruppert (1982b) bounded the influence of the residuals but not leverage. This is acceptable for designed experiments, but not in general.

Example 6.1   continued

We applied the following robust pseudo-likelihood method to the Skeena River data. The model again was that the recruitment variance is proportional to a power of the mean. The algorithm is as follows.

Algorithm for robust pseudo-likelihood method

Step 1   Fix $\hat{\beta}_p$ and $\hat{\theta}_p$ equal to the maximum-likelihood estimates.

Step 2   Fix β in $\mu_i(\beta)$ equal to $\hat{\beta}_p$, and fix θ equal to $\hat{\theta}_p$. (In this example, the variance did not depend on β so that μ_i was not actually needed. We are describing the general algorithm so that it can be applied to other data sets.) Substitute these into the loglikelihood given by equation (6.25). The value of β in $f(x, \beta)$ is allowed to vary. Call the result $\Psi_i(y, \beta)$. As described in section 6.3, calculate the one-step redescending estimator using this Ψ_i. Call this $\hat{\beta}$.

Step 3   Replace $\hat{\beta}_p$ by $\hat{\beta}$.

Step 4   Fix β equal to $\hat{\beta}_p$ in (6.25) and call the result $\Psi_i(y, \theta)$. Again using the techniques of section 6.3, calculate the one-step redescending estimate of θ. Call the estimate $\hat{\theta}$.

Step 5   Replace $\hat{\theta}_p$ by $\hat{\theta}$.

Step 6   Return to step 2.
          End

The algorithm can be checked for convergence at steps 3 and 5. In the case of the Skeena River data, $\hat{\beta}$ on the second iteration was very close to $\hat{\beta}$ from the first iteration, so the algorithm could have been stopped at step 3 of the second iteration. As a check we completed step 5 of the second iteration. The new estimate of θ was virtually the same as on the previous iteration.

The estimates and their standard errors are given in Table 6.1. Although the estimates given above do not agree exactly with those given in Table 6.1 for simultaneous estimation of β and θ, the differences are always quite small compared to the standard errors.

In Table 6.2 the local influence diagnostics are given for separate estimation of β and α. In general, they are similar to the local influence diagnostics when β and α are simultaneously estimated.

For further examples of robust estimation of weighted regression models, the reader is referred to Giltinan et al. (1986).

6.5 Transformation and weighting

In this section we discuss diagnostics and robust estimation for the transformation model with heteroscedastic errors. This model is given in Chapter 5 as equation (5.1), which we restate here

$$h(y_i, \lambda) = h[f(x_i, \beta), \lambda] + \sigma g(\mu_i(\beta), z_i, \theta)\varepsilon_i \qquad (6.29)$$

Although, model (6.29) includes the transform-both-sides model as a special case, for pedagogic reasons we introduced the transform-both-sides model first in Chapter 4. Now we will only treat the general model given by (6.29).

As in section 5.2, let

$$r_i(\beta, \lambda, \theta) = \{h(y_i, \lambda) - h[f(x_i, \beta), \lambda]\}/g(\mu_i(\beta), z_i, \theta)$$

The loglikelihood of y_i given x_i and z_i is

$$l_i(y_i, \beta, \lambda, \theta) = -\log[\sigma(\beta, \lambda, \theta)] - r_i^2/[2\sigma^2(\beta, \lambda, \theta)]$$
$$- \log[g(\mu_i(\beta), z_i, \theta)] + \log[J_i(\lambda)]$$

where $J_i(\lambda)$ is the Jacobian of the transformation from y_i to $h(y_i, \lambda)$. As in section 6.4, the nuisance parameter is eliminated by substituting $\sigma^2(\beta, \lambda, \theta)$, the weighted average of $r_1^2, \ldots, r_N^2$ with weights $w_1, \ldots, w_N$.

The maximum-likelihood estimator maximizes

$$\sum_{i=1}^{N} l_i(y_i, \beta, \lambda, \theta)$$

and it is shown in section 5.2 that the MLE minimizes

$$\sum_{i=1}^{N} e_i^2$$

where

$$e_i = r_i(\beta, \lambda, \theta)\dot{g}(\beta, \theta)/\dot{J}(\lambda)$$

Here $\dot{g}(\beta, \theta)$ and $\dot{J}(\lambda)$ are the geometric means of $\{g(\mu_i(\beta), z_i, \theta)\}$ and $\{J_i(\lambda)\}$, respectively.

The maximum-likelihood estimator is an M-estimator with 'psi function' Ψ_i given by either

$$\nabla l_i(y, \beta, \lambda, \theta) \qquad (6.30)$$

or

$$e_i(\beta, \lambda, \theta)\nabla e_i(\beta, \lambda, \theta) \qquad (6.31)$$

Here ∇ is the gradient with respect to (β, λ, θ). If form (6.31) is used, then the geometric means should be replaced by weighted geometric means, i.e.

$$\log[\dot{J}(\lambda)] = \left(\sum_{i=1}^{N} w_i \log[J_i(\lambda)] \right) \Big/ \left(\sum_{i=1}^{N} w_i \right)$$

In the following example we used (6.30).

Example 6.2   Bacterial clearance

This is a continuation of Examples 4.3 and 5.1. We now return to the bacterial clearance data given in Table 4.6. We will use the model (4.45) for the median response and the variance 'model A' discussed in section 5.3. Since each treatment and time of sacrifice combination is replicated six times, one might look for outliers by informal data analysis such as boxplots. However, we have already seen that such techniques are inadequate unless they account for the skewness and heteroscedasticity of the data. Instead we will use robust estimation, diagnostics, and analysis of the residuals from the robust fit to (5.1) or (6.29).

First we will use the modified power transformation with shift μ fixed at -0.05, i.e., we add 0.05 to the response and the model before power-transforming. Then $\mu = -1$ will be used for comparison.

For simplicity we will concentrate on the bounded-influence estimator. The redescending estimator leads to a similar analysis.

The $\hat{\alpha}$, $\hat{\beta}$, and $\hat{\Delta}$ values in model (4.45) are virtually identical for the MLE and the robust estimators. Therefore we will begin by discussing influence for the transformation parameter λ and the variance parameter π. Table 6.3 gives the MLE, the BI estimator, and the redescending estimator of λ and π for μ fixed at -0.05 or -1. Notice that $\hat{\lambda}$ and $\hat{\pi}$ are somewhat different for the MLE and the robust estimates. However, the choice of μ affects $\hat{\lambda}$ and $\hat{\pi}$ more than the choice of estimator. Also, going from the MLE to the robust estimates has the similar effects on $\hat{\lambda}$ and $\hat{\pi}$ as changing μ from -0.05 to -1.

Table 6.3 *Bacterial clearance data. Estimates of λ and π. The estimators are the maximum-likelihood estimator (MLE), the bounded-influence estimator (BI est.), and the redescending estimator (Redes. est.). Standard errors are given in parentheses*

	$\mu = -0.05$			$\mu = -1$		
	MLE	BI est.	Redes. est.	MLE	BI est.	Redes. est.
$\hat{\lambda}$	−0.069	−0.101	−0.14	−0.18	−0.20	−0.21
	(0.033)	(0.071)	(0.055)	(0.057)	(0.063)	(0.064)
$\hat{\pi}$	3.51	3.82	6.17	5.31	6.98	9.17
	(0.85)	(1.84)	(3.25)	(2.38)	(3.85)	(4.55)

The reason for this will be clear when we examine the local influence diagnostics.

Nine cases deserve further examination. These are listed in Table 6.4 with the following statistics: treatment group, time of sacrifice, response, residual from the BI estimate (standardized by the median absolute deviation, i.e., the MAD), log (estimated median response $-\mu$), and the local influence (LI) diagnostics $\Delta\lambda$ and $\Delta\pi$ evaluated at the BI estimate.

The most influential observation is case 49. This case has a response equal to zero, and power transformation induces a severe outlier even with 0.05 added. Note that $\Delta\lambda$ is positive and $\Delta\pi$ is negative for this observation. Increasing its weight increases $\hat{\lambda}$ and decreases $\hat{\pi}$. This is sensible. Case 49 is an extreme negative residual, which induces left-skewness to the residuals and therefore tends to increase $\hat{\lambda}$. Also, since the time of sacrifice is $t = 24$, this case suggests greater heteroscedasticity, i.e., a more rapid increase in the variance as a function of t. Increased heteroscedasticity is equivalent to a smaller value of π; see model A given by equation (5.8).

Case 72 is also outlying with a negative residual and time of sacrifice equal to 24 h. Like case 49 it is influential for $\hat{\pi}$. However, case 72 has relatively little influence on $\hat{\lambda}$. The reason is that case 72 has a much higher response than case 49, which has a response equal to 0. Unless μ is large (say 0.5 or greater) the shifted response of case 49 will be close to the singularity of the negative power transformations occurring at 0. Therefore, case 49 prevents $\hat{\lambda}$ from decreasing much below 0. The robust estimators downweight case 49 and this decreases $\hat{\lambda}$. Changing μ to -1 has virtually the same effect as using a robust estimator, because using $\mu = -1$ keeps the shifted response of case 49 away from 0.

In Table 6.4 we see that if μ is changed from -0.05 to -1, then the influence of case 49 on $\hat{\lambda}$ becomes quite small, but its influence on $\hat{\pi}$ increases.

Some cases, in particular cases 19 and 48 and to a lesser extent cases 4, 11, and 34, are downweighted by the BI estimator even though they have little influence on $\hat{\lambda}$ and $\hat{\pi}$. These cases have large effects on the estimated α, β, and Δ values, large at least relative to the standard errors of these estimates. These effects are seen from the local influence diagnostics. We will not give the LI diagnostics here but will summarize the results. Case 4 has a standardized residual equal to -2 and occurs at $t = 0$. It is in group O (control) and decreases $\hat{\alpha}_1$.

Table 6.4 Bacterial clearance data. Diagnostics for selected cases. Log(m) is the log of the estimate of the shifted median. For each case the first row is for $\mu = -0.05$ and the second row is for $\mu = -1$. The residuals, case weights, and the estimated medians use the bounded-influence estimator. $\Delta\lambda$ and $\Delta\pi$ are the local influence diagnostics evaluated at the bounded-influence estimator

Case	Group	Time	Response	Residual	Weight	log(m)	Δλ	Δπ
4	O	0	41.4	-2	0.51	4.33	0.005	0.619
				-1.80	0.64	4.31	0	0.657
11	A	0	42.8	-1.8	0.64	4.29	-0.003	0.457
				-1.6	0.72	4.28	0	0.498
19	VA	0	128	0.94	0.24	4.55	0	0.041
				0.83	0.30	4.55	0	-0.47
34	A	4	48.7	2	0.74	3.06	0.009	-0.083
				2.5	0.71	3.04	0.013	0.422
48	VA	4	112.4	2.7	0.44	3.50	0.008	-0.197
				3.2	0.47	3.49	0.005	0.827
49	O	24	0	-2.8	0.23	0.115	0.138	-1.73
				-1.4	0.79	0.136	0.010	-0.392
57	A	24	0.2	-1.1	0.67	-0.096	0.007	-0.115
				-0.9	0.96	-0.021	-0.002	0.033
64	V	24	211	-1.4	1.00	9.18	-0.016	-0.164
				-1.7	0.91	8.97	-0.002	-0.644
72	VA	24	3.3	-2.3	0.43	5.28	0.01	-1.00
				-3.9	0.28	5.27	0.037	-6.416

Case 11 is similar to case 4 but is in group A (ampicillin) and decreases $\hat{\alpha}_2$.

Cases 19 and 48 have positive residuals and are in group VA (virus). Case 19 is at $t = 0$ and decreases $\hat{\beta}_4$ while increasing $\hat{\Delta}_4$. Case 48 is at $t = 4$ and has the opposite effects on these estimates.

The main point of the diagnostics is that they show how sensitive the estimated power transformation $\hat{\lambda}$ is to the shift μ when the responses include a zero value. This sensitivity could be expected but its magnitude came as a surprise to us. It is interesting and somewhat unexpected that $\hat{\pi}$ is also very sensitive to the choice of μ. The diagnostics also show how certain cases change the estimates of parameters in the model for the median response, but for this data set these changes are of secondary importance.

Perhaps μ should be estimated from the data. However, as Atkinson (1985) discusses in detail, μ cannot be estimated by normal-theory maximum likelihood. As the minimum of the shifted responses, $y_i - \mu$, tends to zero, the likelihood tends to infinity. This should not be viewed as failure of the method of maximum likelihood. The problem is that the transformation-to-normality model, which can never hold exactly, cannot even hold approximately when the shifted responses are near 0. The skewness and heteroscedasticity estimators, $\hat{\lambda}_{sk}$ and $\hat{\lambda}_{het}$, introduced in section 4.3, may be promising in this context, since they do not depend on the normality assumption.

We now examine the residuals more closely to see if the transformation and variance model A really has achieved near-normality and a constant variance. We will use the residuals from the redescending estimate and $\mu = -0.05$. The redescending estimate makes outliers stand out, and with this estimate the choice of μ is not crucial.

Figure 6.1 is a plot of the residuals (standardized by the MAD) against the logarithms of the fitted values. This graph should be compared to Figure 5.2, which is based on the same model but uses the MLE. Recall that in Figure 5.2 there seemed to be too little variability among the cluster of 12 observations with small fitted values. This has changed in Figure 6.1. Because $\hat{\lambda}$ changes from -0.07 to -0.14 as we change from the MLE to the redescending estimate, cases 57 and 49, particularly the latter, are negative outliers in Figure 6.1. Inducing negative outliers has achieved a more nearly constant variance, but this is not a very satisfactory way of obtaining homoscedasticity.

Figure 6.2 is a normal plot of the residuals. Cases 49, 72, 34, and 48 are clearly outlying, but the remaining data are approximately

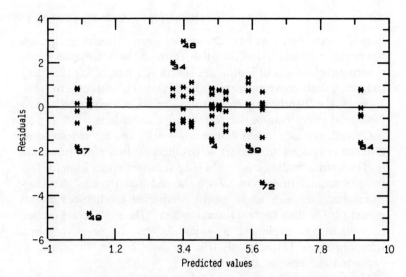

Figure 6.1 *Bacterial clearance data. Residuals and predicted values. Residuals have been standardized by the MAD. Transform both sides with variance model A. Redescending estimator.* $\mu = -0.05$. *Selected cases indicated.*

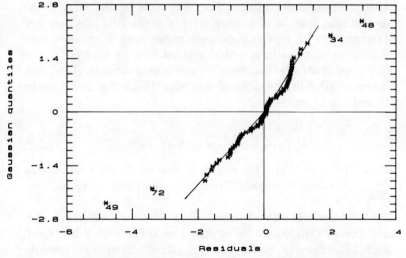

Figure 6.2 *Bacterial clearance data. Normal probability plot of residuals. Residuals have been standardized by the MAD. Transform both sides with variance model A. Redescending estimator.* $\mu = -0.05$. *Selected cases indicated.*

normal. In the figure we have drawn two straight lines, one through the negative residuals and the other through the positive residuals, ignoring the outliers in both cases. There is a hint of left-skewness, indicating slight overtransformation. This is not of real concern, but it supports the slightly weaker log-transformation, $\lambda = 0$, which has a biological interpretation (see section 5.1 and below).

Overall, variance model A with $\mu = -0.05$ and the redescending estimate provides a satisfactory fit, but there are two minor problems. (1) The transformation $\hat{\lambda} = -0.14$ is 2.5 standard errors from the log-transformation. In section 5.3 we showed that the unshifted log-transformation leads to an easily interpreted random-coefficients model. (2) The data contain several outliers. The most severe outlier, case 49, can be explained; it is induced by the negative power transformation. This is partly true for case 72. Case 48 cannot be explained as transformation-induced.

Although we could stop our analysis here, there is an interesting hypothesis worth pursuing. Recall that variance model A was developed by assuming that the bacterial growth rate in the lung varied between animals. Until now we have assumed that the variance of the growth rate does not depend on the treatment group. However, viral exposure could add another component of variability to the growth rate, because of varying severity of the viral infection. For example, case 72 is a viral-exposed mouse with a response only slightly higher than the non-viral-infected mice. Perhaps this mouse had a low-grade viral infection. We now extend the model to allow a different variability for virus-infected mice. This will be called model C, and can be written

$$\sigma_{itj} = \sigma g_C(t, \pi_1, \pi_2)$$
$$= \sigma\{\pi_1^2 + (1 + \pi_2^2 I[i = V \text{ or } VA])(t + 0.25)^2\}^{1/2} \quad (6.32)$$

Here $I[i = V \text{ or } VA]$ is the indicator function for the viral-infected groups. If $\pi_2 = 0$, then model C reduces to model A given by equation (5.8).

Model C was fit with μ fixed at -0.05. First λ was allowed to vary and then λ was fixed at 0. The estimates are in Table 6.5. With λ fixed, the MLE of π_1 and π_2 are both about 2.5 standard errors away from 0. When λ varies, the standard errors increase considerably, and neither π_1 nor π_2 are significantly different from 0 at $p = 0.05$. However, with λ allowed to vary, the loglikelihood is -283.6 for model C and -286.1 for model A. Twice the difference is 5.0, which is significant at

Table 6.5 *Bacterial clearance data. Estimates of transformation and variance parameters. Model C, the two-parameter variance model, $\mu = -0.05$. Standard errors are in parentheses*

Estimator	λ fixed		λ not fixed		
	$\hat{\pi}_1$	$\hat{\pi}_2$	$\hat{\lambda}$	$\hat{\pi}_1$	$\hat{\pi}_2$
MLE	6.42	2.03	0.060	9.31	3.55
	(2.45)	(0.83)	(0.049)	(5.07)	(2.40)
BI est.	9.50	2.96	0.048	12.96	4.53
	(4.93)	(1.66)	(0.056)	(5.37)	(2.68)
Redes. est.	20.2	5.89	0.007	21.5	6.06
	(10.1)	(3.11)	(0.044)	(12.5)	(4.81)

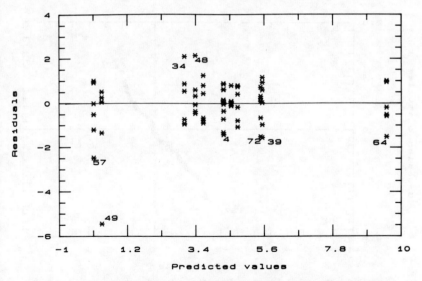

Figure 6.3 *Bacterial clearance data. Residuals and predicted values. Transform both sides with variance model C. Residuals have been standardized by the M AD. Redescending estimator. $\lambda = \hat{\lambda} = 0.007$. $\mu = -0.05$. Selected cases indicated.*

$p = 0.05$. Thus model C is supported by the data, at least if we retain the most influential observations. However, it is difficult to estimate λ, π_1, and π_2 simultaneously; their estimators are highly correlated with large standard errors for $\hat{\pi}_1$ and $\hat{\pi}_2$. This suggests fixing λ at 0, a value supported by biological reasoning.

The residual plots for model C using the redescending estimate are given as Figures 6.3 and 6.4. These plots are with $\lambda = \hat{\lambda} = 0.044$, but the plots with $\lambda = 0$ are nearly the same. There is little indication of heteroscedasticity in Figure 6.3 or in plots (not shown) of the absolute residuals against time and against the logarithms of the fitted values. Notice that case 72 is no longer outlying and case 48 now has a residual of only 2.1, though the residual of case 49 is an extreme outlier, -5.5. Model C seems preferable to model A because it 'explains' the outliers, except for the transformation-induced outlier, case 49. The low variability for groups V and VA at $t = 24$ is 'explained' by the variance model rather than removed by a negative power transformation as when model A is fit. Therefore, case 72 is no longer a transformation-induced outlier, and case 48 is now seen as less outlying since it belongs to a viral-infected group.

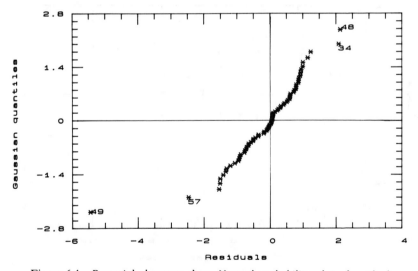

Figure 6.4   *Bacterial clearance data. Normal probability plot of residuals. Residuals have been standardized by the MAD. Transform both sides with variance model C. Redescending estimator.* $\lambda = \hat{\lambda} = 0.044$. $\mu = -0.05$. *Selected cases indicated.*

Table 6.6   *Bacterial clearance data. Model C.* $\mu = -0.05$. *Case weights and LI diagnostics evaluated at the redescending estimate*

	λ not fixed				λ fixed		
		LI diagnostic				LI diagnostic	
Case	Weight	$\Delta\lambda$	$\Delta\pi_1$	$\Delta\pi_2$	Weight	$\Delta\pi_1$	$\Delta\pi_2$
4	0.93	−0.01	1.23	−0.07	0.86	1.77	0.25
11	0.78	−0.01	0.78	−0.12	0.71	1.23	0.14
19	0	0	0.08	0.01	0	0.08	0.01
34	0.55	0	2.95	0.26	0.54	2.56	0.11
48	0.006	0.05	5.19	3.26	0.005	1.31	1.04
49	0	0.54	−74.6	−10.1	0	−114	−34.1
57	0	0.07	−16.4	−3.19	0	−21.1	−6.32
64	0.49	−0.02	−1.58	−0.15	0.86	−0.01	0.47
72	0.006	0.03	2.26	2.03	0.51	−0.08	0.57

Notice in Table 6.5 that $\hat{\pi}_1$ and $\hat{\pi}_2$ have much larger standard errors when using the redescending estimate rather than the MLE. This is because $\hat{\pi}_1$ and $\hat{\pi}_2$ rely on information from case 49 which is given weight 0 by the redescending estimate.

Table 6.6 gives diagnostics. Only case 49 has much influence on $\hat{\lambda}$. It might be that this is the case that drives $\hat{\lambda}$ positive; notice that $\hat{\lambda}$ decreases to nearly 0 as one switches from the MLE to the redescending estimate. Numerous cases influence $\hat{\pi}_1$ and $\hat{\pi}_2$, though case 49 has the largest effects. Clearly π_1 and π_2 are not estimated very precisely, and this is reflected in their standard errors.

We prefer model C to model A. Adding the extra parameter 'explains' all but one of the outliers. Although adding parameters to fit outliers should *not* be routine practice, in this case the added parameter has a biological interpretation and supports the log-transformation, which is also based on a biological theory.

The goal of regression analysis is not merely to find a single estimate but rather to understand the data, and more importantly the phenomena being studied. However, one usually does end the analysis by accepting one model and estimate, even if the acceptance is only tentative and will change with the input of new data or new scientific theory. Therefore, we propose model C with $\mu = -0.05$,

$\lambda = 0$, and the redescending estimate. One needs the robust estimate to deal with case 49. One could use $\mu = -1$ to stablize case 49, but we prefer keeping μ close to 0 because the unshifted log-transformation has a neat biological interpretation.

It is interesting that cases 62 and 63, which are severe positive outliers before transformation, fit the transformation model quite well. Before beginning the analysis we expected them to be the most influential cases. The most influential case, case 49, looks innocuous before transformation.

CHAPTER 7

Technical complements

7.1 M-estimators

This section briefly reviews the theory of M-estimators for independent but not necessarily identically distributed observations. We assume that $\{y_i\}_{i=1}^N$ is a collection of independent random variables and that y_i has distribution F_i. A special case of much importance is when the observations are identically distributed so that $F_i = F$ for some F and all i.

Let θ be a parameter depending on $\{F_i\}_{i=1}^N$ and taking values in the parameter space Θ which is an open subset of $\mathbb{R}^p$, $p \geqslant 1$. We will distinguish between parametric and nonparametric estimation. In the former case we have a parametric family of distributions $F_i(y, t)$, $t \in \Theta$ and $i = 1, \ldots, N$, and for some θ (independent of i) $F_i = F_i(y, \theta)$ for all i. Thus F_i is assumed known except for a p-dimensional parameter. In the nonparametric case, θ is determined by $\{F_i\}_{i=1}^N$ but not vice versa. We also assume that for some measure m, $F_i(\cdot, t)$ has a density $f_i(\cdot, t)$ with respect to m. Linear models with Gaussian errors and a constant or more generally parametric variance function are parametric problems, while linear models with arbitrary mean-zero errors are nonparametric problems.

M-estimators can be defined in two ways: (a) as the solution of a minimization problem and (b) as the root of an estimating equation. In case (a) $\hat{\theta}$ is defined by

$$\sum_{i=1}^N \rho_i(y_i, \hat{\theta}) = \min! \tag{7.1}$$

that is, $\hat{\theta}$ minimizes the sum on the left-hand side of (7.1), where ρ is a real-valued function. In case (b) define $\hat{\theta}$ so that

$$\sum_{i=1}^N \Psi_i(y_i, \hat{\theta}) = 0 \tag{7.2}$$

In both (7.1) and (7.2) the existence and uniqueness of $\hat{\theta}$ are important questions but are difficult to answer except in special cases. If $\rho_i(y, \cdot)$ is differentiable for each fixed y then (7.1) implies (7.2) with $\Psi_i(y, t) = (\partial/\partial\theta)\rho_i(y, \theta)$, but of course (7.2) does not imply (7.1) since (7.2) is satisfied by local minima or even maxima, if they exist. Consequently, it is preferable if an estimator can be defined by (7.1). However, a p-dimensional function Ψ_i may not be the gradient of any real-valued function ρ_i. Therefore, when $p > 1$ then the class of estimators defined by estimating equations of the form (7.2) is more general than the class defined by (7.1). In the identically distributed case we assume that for some Ψ, $\Psi_i = \Psi$ for all i.

Maximum-likelihood estimators are M-estimators, and in fact 'M-estimator' was coined by Huber (1964) to mean of maximum-likelihood type. Let $g_i(y, t)$, $t \in \Theta$, be a family of densities and let

$$s_i(y, t) = (\partial/\partial t)\log g_i(y, t)$$

be the so-called 'score function'. Setting $\rho_i(y, t) = -\log g_i(y, t)$ in (7.1) or $\Psi_i = s_i$ in (7.2) gives the MLE.

For parametric estimation problems, we say that we are 'at the ideal model' when

$$g_i(y, t) = f_i(y, t)$$

that is, the density of the data is in the same family that defines the estimator. Maximum-likelihood estimation is also used when the F_i are not in a parametric family. For example, the least squares estimator for a regression model is the MLE if the errors are independent, identically distributed $N(0, \sigma^2)$, but the least-squares estimator is often used when this assumption is not met. The maximum-likelihood estimator is asymptotically efficient at the ideal model under reasonable regularity conditions, but can be very inefficient off the ideal model. This point is discussed in more detail in Chapter 6 and the references given there. Certain modifications of the maximum-likelihood estimator, which are M-estimators with Ψ_i a suitably truncated version of the score function, retain their good efficiency when the data are only close to the ideal model. These robust alternatives to the MLE are introduced in Chapter 6.

7.1.1 Consistency

An assumption that is quite close to being necessary for the consistency of $\hat{\theta}$ is:

Assumption 1:   Case (1): The function

$$\sum_{i=1}^{N} \int \rho_i(y, t) dF_i(y) \tag{7.3}$$

has a unique minimum at $t = \theta$, or
Case (b): The equation

$$\sum_{i=1}^{N} \int \Psi_i(y, t) dF_i(y) = 0 \tag{7.4}$$

has a unique solution $t = \theta$. In the independent, identically distributed case (7.4) is equivalent to

$$\int \Psi(y, t) dF(y) = 0 \tag{7.5}$$

In the case of maximum likelihood estimation, (7.3) becomes

$$\sum_{i=1}^{N} \int \log f_i(y, t) dF_i(y) = \max!$$

It is well-known that

$$\int \log f_i(y, t) f_i(y, \theta) d\mu(y)$$

is uniquely maximized by $t = \theta$, provided that for all $t \neq t'$ $f_i(\cdot, t)$ and $f_i(\cdot, t')$ are distinct densities. Thus assumption 1 holds for maximum likelihood estimation at the ideal model. Off the model, the MLE estimates the value of $t \in \theta$ that minimizes the Kullback–Leibler discrepancy between the data density and the parametric model; see Hernandez and Johnson (1980) who discuss Box–Cox transformations when no transformation to exact normality is possible.

Of course assumption 1 alone will not guarantee consistency without additional regularity conditions. Moreover, it seems possible to construct pathological examples where assumption 1 is violated mildly but $\hat{\theta} \to \theta$ nonetheless. Huber (1967) gives a rigorous discussion of consistency for independent, identically distributed data.

7.1.2 Asymptotic normality

Once consistency has been established, the asymptotic distribution of $\hat{\theta}$ can be found by the following heuristic argument, which can be made rigorous in many important special cases. Let $\dot{\Psi}_i(y,t) = \partial/\partial t \Psi(y,t)$. By a Taylor series argument

$$0 = \sum_{i=1}^{N} \Psi_i(y_i, \hat{\theta})$$

$$\cong \sum_{i=1}^{N} \Psi_i(y_i, \theta) + \left[\sum_{i=1}^{N} \dot{\Psi}_i(y_i, \theta) \right](\hat{\theta} - \theta),$$

so that

$$N^{1/2}(\hat{\theta} - \theta) \cong - B_N^{-1} \left[N^{-1/2} \sum_{i=1}^{N} \Psi_i(y_i, \theta) \right], \tag{7.6}$$

where

$$B_N = E\left[N^{-1} \sum_{i=1}^{N} \dot{\Psi}_i(y_i, \theta) \right] \tag{7.7}$$

We assume that B_N converges as $N \to \infty$ to some nonsingular matrix B. We also assume that

$$N^{-1/2} \sum_{i=1}^{N} \Psi_i(y_i, \theta) \to N(0, A) \tag{7.8}$$

for some matrix A. In the identically distributed case we assume

$$A = \int \Psi(y, \theta) \Psi(y, \theta)^{\mathrm{T}} dF(y)$$

is finite. Define

$$A_N = E\left[N^{-1} \sum_{i=1}^{N} \Psi_i(y_i, \theta) \Psi_i^{\mathrm{T}}(y_i, \theta) \right].$$

In the non-identically distributed case, (7.8) will hold if $A_N \to A$ and other regularity conditions hold. By (7.6)–(7.8) we have

$$N^{1/2}(\hat{\theta} - \theta) \xrightarrow{D} N(0, (B^{-1}AB^{-\mathrm{T}})), \tag{7.9}$$

where we use the notation $M^{-\mathrm{T}} = (M^{-1})^{\mathrm{T}}$.

For parametric estimation we have a useful identity for B_N:

$$B_N = - N^{-1} \sum_{i=1}^{N} \int \Psi_i(y, \theta) s_i(y, \theta)^{\mathrm{T}} dF_i(y) \tag{7.10}$$

which follows from differentiating the identity

$$\sum_{i=1}^{N} \int [\Psi_i(y,t) f_i(y,t)] dm(y) \equiv 0,$$

and setting t equal to θ. In the case of maximum likelihood estimation $-B_N$ is called the Fisher information matrix and will be denoted by I_θ. Then since $\Psi_i = s_i$, (7.10) becomes the well-known identity

$$I_\theta = N^{-1} \sum_{i=1}^{N} \int s_i(y, \theta) s_i(y, \theta)^{\mathrm{T}} dF_i(y) = A_N$$

so that in (7.9) we have $(B_N^{-1} A_N B_N^{-\mathrm{T}}) = I_\theta^{-1}$. We can estimate A_N and B_N by

$$\hat{A} = N^{-1} \sum_{i=1}^{N} \Psi_i(y_i, \hat{\theta}) \Psi_i(y_i, \hat{\theta})^{\mathrm{T}} \tag{7.11}$$

and

$$\hat{B} = N^{-1} \sum_{i=1}^{N} \dot{\Psi}_i(y_i, \hat{\theta}) \tag{7.12}$$

or

$$B^* = N^{-1} \sum_{i=1}^{N} \Psi_i(y_i, \hat{\theta}) s_i^{\mathrm{T}}(y_i, \hat{\theta}). \tag{7.13}$$

The approximation

$$\hat{\theta} \cong N(\theta, N^{-1} \hat{B}^{-1} \hat{A} \hat{B}^{-\mathrm{T}}) \tag{7.14}$$

or

$$\hat{\theta} \cong N(\theta, N^{-1} (B^*)^{-1} \hat{A} (B^*)^{-\mathrm{T}}) \tag{7.15}$$

can be used for large-sample tests or confidence regions. Often $\dot{\Psi}_i$ is difficult to calculate due to the complexity of Ψ_i and therefore (7.15) is easier to use. However, (7.14) has the advantage of being nonparametric. Since (7.14) does not depend on any parametric assumptions it can be used with maximum likelihood estimators when F_i is not necessarily a member of the parametric family $F_i(\cdot, t)$, $t \in \Theta$.

7.2 Wald, score, and likelihood ratio tests

Throughout this section, the parameter θ will be a vector of dimension p, split into the components $\theta = (\gamma, \lambda)$, where γ has dimension $q \leqslant p$. We are interested in testing the hypothesis

$$H_0: \gamma = \gamma_0$$

The loglikelihood based on N observations will be denoted $L(\gamma, \lambda)$.

7.2.1 Wald tests

The original meaning of the term 'Wald test' is based on the work of

Wald (1943) (see also Rao, 1973). We use the term in a somewhat more general fashion. As discussed in sections 2.5 and 4.3, Wald tests are easy to construct, but due to inadequacy of the asymptotic approximations their levels may not be accurate. The major advantage of Wald tests is their ease of construction and generality. For example, Wald tests can be constructed for M-estimators satisfying either (7.14) or (7.15). Suppose that $\hat{\gamma}$ is asymptotically normally distributed with mean γ and covariance matrix Ω/N. In symbols,

$$N^{1/2}(\hat{\gamma} - \gamma) \to N(0, \Omega)$$

For M-estimators, this is the upper-left $q \times q$ submatrix of $(B^{-1}AB^{-1})$ given in (7.9). Let $\hat{\Omega}$ be a consistent estimate of Ω, i.e.

$$\hat{\Omega} \xrightarrow{p} \Omega$$

For M-estimators, see (7.11)–(7.13). The Wald test statistic is

$$N(\hat{\gamma} - \gamma_0)^T \hat{\Omega}^{-1}(\hat{\gamma} - \gamma_0) = T_W(\gamma_0) \qquad (7.16)$$

Under the hypothesis, the test statistic (7.16) has an asymptotic chi-squared distribution with q degrees of freedom. The test with asymptotic level α rejects the hypothesis if (7.16) exceeds the appropriate percentile of the chi-squared distribution, i.e.

$$\text{Reject if } T_W(\gamma_0) \geq \chi^2(1 - \alpha, q) \qquad (7.17)$$

Because of the analogy with linear regression, the chi-squared percentile in (7.17) is often replaced by the $1 - \alpha$ percentile of the F-distribution with q and $N - p$ degrees of freedom.

If γ and hence Ω are scalars, under the hypothesis

$$t_W(\gamma_0) = N^{1/2}(\hat{\gamma} - \gamma_0)/\hat{\Omega}^{1/2} \to N(0, 1)$$

Again, in analogy with linear regression the usual practice is to approximate the distribution of the test statistic $t_W(\gamma_0)$ by a t-distribution with $N - p$ degrees of freedom, and to reject the hypothesis if

$$|t_W(\gamma_0)| \geq t(1 - \alpha, N - p)$$

7.2.2 Likelihood ratio tests

Let $L(\hat{\gamma}, \lambda)$ be the log-likelihood function based on N observations. Define $(\hat{\gamma}, \lambda)$ as the maximum-likelihood estimate of (γ, λ), and let λ_0 be the maximum-likelihood estimate of λ constraining $\gamma = \gamma_0$. The likelihood ratio test statistic is

$$T_L(\gamma_0) = 2[L(\hat{\gamma}, \hat{\lambda}) - L(\gamma_0, \hat{\lambda}_0)] \qquad (7.18)$$

Under the hypothesis, this statistic has an asymptotic chi-squared distribution with q degrees of freedom, and the test is

$$\text{Reject } H_0 \text{ if } T_L(\gamma_0) \geqslant \chi^2(1 - \alpha, q)$$

Generally, in the test statistic (7.18), maximum-likelihood estimates must be used or the large-sample null distribution will not be chi-squared.

7.2.3   Score tests

The score test when there are nuisance parameters is defined in Cox and Hinkley (1974, p. 324). Let $\tilde{\lambda}$ be any $N^{1/2}$-consistent estimate of λ; this could but need not necessarily be the maximum-likelihood estimate under the constraint $\gamma = \gamma_0$. Define

$$\iota_{\gamma\lambda} = -E\left(\frac{\partial^2}{\partial\gamma\,\partial\lambda^T} L(\gamma, \lambda) \right)$$

$$\iota_{\lambda\lambda} = -E\left(\frac{\partial^2}{\partial\gamma\,\partial\gamma^T} L(\gamma, \lambda) \right)$$

$$\iota_{\lambda\lambda} = -E\left(\frac{\partial^2}{\partial\lambda\,\partial\lambda^T} L(\gamma, \lambda) \right)$$

Let $\iota^{\gamma\gamma}$ be the upper left-hand corner of the matrix

$$\iota^{-1} = \begin{bmatrix} \iota_{\gamma\gamma} & \iota_{\gamma\lambda} \\ \iota_{\gamma\lambda}^T & \iota_{\lambda\lambda} \end{bmatrix}^{-1}$$

i.e.

$$\iota^{\gamma\gamma} = (\iota_{\gamma\gamma} - \iota_{\gamma\lambda}\iota_{\lambda\lambda}^{-1}\iota_{\gamma\lambda}^T)^{-1}$$

The score test statistic for H_0: $\gamma = \gamma_0$ is given by

$$T_S(\gamma_0, \hat{\lambda}) = Q(\gamma_0, \hat{\lambda})^T \iota^{\gamma\gamma}(\gamma_0, \hat{\lambda}) Q(\gamma_0, \hat{\lambda})$$

where

$$Q(\gamma, \lambda) = L_\gamma(\gamma, \lambda) - \iota_{\gamma\lambda}(\gamma, \lambda)\iota_{\lambda\lambda}^{-1}(\gamma, \lambda)L_\lambda(\gamma, \lambda)$$

If $\hat{\lambda}$ is the maximum-likelihood estimate of λ under the constraint that $\gamma = \gamma_0$, then $L_\lambda(\gamma_0, \hat{\lambda}) = 0$ and the test statistic takes the particularly simple form

$$T_S(\gamma_0, \hat{\lambda}) = L_\gamma(\gamma_0, \hat{\lambda})^T i^{\gamma\gamma}(\gamma_0, \hat{\lambda}) L_\gamma(\gamma_0, \hat{\lambda})$$

Under the hypothesis, T_S has an asymptotic chi-squared distribution with q degrees of freedom and the test is

$$\text{Reject } H_0 \text{ if } T_S(\gamma_0, \hat{\lambda}) \geqslant \chi^2(1 - \alpha, q)$$

7.3 Miscellaneous technical results

7.3.1 Sketch of the proof of Theorem 2.1

We sketch a proof of a more general result than Theorem 2.1.

Theorem Suppose that the vector Y has mean vector $\mu(\beta)$, where

$$\mu^T = (\mu_1, \mu_2, \ldots, \mu_N)$$

and $\mu_i = f(x_i, \beta)$. Let the covariance of Y be $\sigma^2 \Lambda(X, Z, \beta, \theta)$, where X is the $(N \times p)$ matrix with rows $\{x_i^T\}$ and Z is an $(N \times q)$ matrix with rows $\{z_i^T\}$. Suppose further that there exist estimates $\hat{\beta}_*$ and $\hat{\theta}$ such that

$$N^{1/2}(\hat{\beta}_* - \beta) = O_{pr}(1) \qquad (7.19)$$

$$N^{1/2}(\hat{\theta} - \theta) = O_{pr}(1) \qquad (7.20)$$

where $O_{pr}(1)$ means bounded in probability. Now form a generalized least-squares estimate $\hat{\beta}_G$ by minimizing

$$[Y - \mu(\beta)]^T \Lambda^{-1}(X, Z, \hat{\beta}_*, \hat{\theta})[Y - \mu(\beta)] \qquad (7.21)$$

Then $\hat{\beta}_G$ is asymptotically normally distributed with mean β and covariance matrix $(\sigma^2/N)S_G^{-1}$, where

$$S_G = \lim_{N \to \infty} N^{-1} D(\beta)^T \Lambda^{-1}(X, Z, \beta, \theta) D(\beta) \qquad (7.22)$$

and $D(\beta)$ is the $(N \times p)$ matrix of partial derivatives of $\mu(\beta)$.

Sketch of the proof Write

$$U(\gamma_1, \gamma_2, \theta) = D(\gamma_1)^T \Lambda^{-1}(X, Z, \gamma_2, \theta)[Y - \mu(\gamma_1)]. \qquad (7.23)$$

Then the generalized least-squares estimate satisfies

$$U(\hat{\beta}_G, \hat{\beta}_*, \hat{\theta}) = 0$$

By a Taylor series expansion, we have that

$$0 \simeq N^{-1/2}U(\beta,\beta,\theta) + N^{-1}\left(\frac{\partial}{\partial\theta}U(\beta,\beta,\theta)\right)N^{1/2}(\hat{\theta}-\theta)$$

$$+ N^{-1}\left(\frac{\partial}{\partial\gamma_2}U(\beta,\gamma_2=\beta,\theta)\right)N^{1/2}(\hat{\beta}_*-\beta)$$

$$+ N^{-1}\left(\frac{\partial}{\partial\gamma_1}U(\gamma_1=\beta,\beta,\theta)\right)N^{1/2}(\hat{\beta}_G-\beta)$$

Since the expected value of $Y-\mu(\beta)$ is zero, an inspection of (7.23) and sufficient regularity conditions show that

$$N^{-1}\frac{\partial}{\partial\theta}U(\beta,\beta,\theta) \xrightarrow{p} 0$$

$$N^{-1}\frac{\partial}{\partial\gamma_2}U(\beta,\gamma_2=\beta,\theta) \xrightarrow{p} 0$$

$$N^{-1}\frac{\partial}{\partial\gamma_1}U(\gamma_1=\beta,\beta,\theta) \xrightarrow{p} -S_G$$

Since $N^{-1/2}U(\beta,\beta,\theta)$ is asymptotically normally distributed with mean zero and covariance matrix $\sigma^2 S_G$, an application of Slutsky's lemma completes the sketch of the proof.

7.3.2 Asymptotic justification of the normal-theory likelihood test

Refer to section 2.5 for the testing problem. With $\beta^T=(\beta_1^T,\beta_2^T)$, we are testing $H_0: \beta_2=\beta_{2,0}$. Define

$$\eta_i^2 = g^2(\mu_i(\hat{\beta}_*),z_i,\theta)$$

$$a_i(\beta) = \frac{\partial}{\partial\beta}f(x_i,\beta)$$

$$c_i(\beta) = \frac{\partial}{\partial\beta_1}f(x_i,\beta)$$

We have that

$$\hat{\sigma}_G^2 = (N-p)^{-1}\sum_{i=1}^N [y_i-f(x_i,\hat{\beta}_G)]^2/\eta_i^2$$

$$\hat{\sigma}_{G,0}^2 = (N-p)^{-1}\sum_{i=1}^N [y_i-f(x_i,\hat{\beta}_{G,0})]^2/\eta_i^2$$

By a Taylor series, the test statistic (2.17) is

$$N[\log(\hat{\sigma}_{G,0}^2) - \log(\hat{\sigma}_G^2)] \simeq N(\hat{\sigma}_{G,0}^2 - \hat{\sigma}_G^2)/\hat{\sigma}_G^2$$
$$+ (N/2)(\hat{\sigma}_{G,0}^2 - \hat{\sigma}_G^2)^2/\sigma_G^4 + o_{pr}(1) \quad (7.24)$$

By standard likelihood calculations and using the fact that the estimates are from generalized least squares we obtain

$$N\left(N^{-1} \sum_{i=1}^N [y_i - f(x_i,\beta)]^2/\eta_i^2 - \hat{\sigma}_G^2\right) \simeq N(\hat{\beta}_G - \beta)^T S_G(\hat{\beta}_G - \beta)$$

$$(7.25)$$

Defining

$$S_{G,0} = N^{-1} \sum_{i=1}^N c_i(\beta)c_i(\beta)^T/g^2(\mu_i(\beta), z_i, \theta)$$

and assuming that $\hat{\beta}_{G,0} - \beta = O_{pr}(N^{-1/2})$, we obtain from a similar calculation that

$$N\left(N^{-1} \sum_{i=1}^N [y_i - f(x_i,\beta)]^2/\eta_i^2 - \hat{\sigma}_{G,0}^2\right) \simeq N(\hat{\beta}_{G,0} - \beta)^T S_{G,0}(\hat{\beta}_{G,0} - \beta)$$

$$(7.26)$$

Combining (7.24)–(7.26), we see that the test statistic (2.17) can be approximated by

$$-(N/\sigma^2)[(\hat{\beta}_{G,0} - \beta)^T S_{G,0}(\hat{\beta}_{G,0} - \beta) - (\hat{\beta}_G - \beta)^T S_G(\hat{\beta}_G - \beta)] \quad (7.27)$$

If we define ε_i as in (2.21), then the proof of Theorem 2.1 given in this section assures us that under the null hypothesis

$$N^{1/2}(\hat{\beta}_G - \beta)/\sigma \simeq S_G^{-1} N^{-1/2} \sum_{i=1}^N a_i(\beta)\varepsilon_i$$

$$N^{1/2}(\hat{\beta}_{G,0} - \beta)/\sigma \simeq S_{G,0}^{-1} N^{-1/2} \sum_{i=1}^N c_i(\beta)\varepsilon_i$$

Standard multivariate normal calculations show that the normal-theory likelihood ratio test has an asymptotic chi-squared distribution under the null hypothesis with $p - r$ degrees of freedom.

The quasi-likelihood test (2.19) can also be written as (7.27) assuming only that $\hat{\beta}_{G,0} - \beta = O_{pr}(N^{-1/2})$, which would assure that the two tests have the same asymptotic level and local power. For example, note from (2.18) that

$$2[L_Q(\beta) - L_Q(\hat{\beta}_G)] \simeq 2 \sum_{i=1}^N l_Q(\mu_i(\hat{\beta}_G), y_i, \theta) f_\beta(x_i, \hat{\beta}_G)^T(\beta - \hat{\beta}_G)$$

$$- N(\beta - \hat{\beta}_G)^T \left(S_G - \sum_{i=1}^{N} [y_i - f(x_i, \beta)] f_{\beta\beta}(x_i, \beta)/g_i^2 \right)(\beta - \hat{\beta}_G)$$

$$\simeq - N(\hat{\beta}_G - \beta)^T S_G(\hat{\beta}_G - \beta)$$

7.3.3 Rothenberg's variance expansion

Suppose that the errors (2.21) are normally distributed and that the variances are not a function of the mean. Rothenberg (1984) uses a clever application of complete sufficiency to show that, the more variable the estimator of θ, the larger the correction term in the second-order variance expansion (3.5):

$$\text{Covariance}[N^{1/2}(\hat{\beta}_G - \beta)] \simeq \sigma^2 S_G^{-1} + N^{-1} V(\hat{\theta})$$

The result holds only for linear regression assuming normal errors. As in the previous chapter, write the model as

$$Y = \mu(\beta) + U = X\beta + U$$

where U is normally distributed with mean zero and covariance matrix

$$\sigma^2 \Lambda(Z, \theta)$$

which does not depend on the mean. Assume that $\hat{\theta}$ when written as a function of $X\beta + U$ does not depend on β and is even in U. Write the weighted least-squares estimator calculated at θ as $\hat{\beta}(\theta)$ and by convention let θ_0 be the true value of θ. Since the observations are normally distributed, $\hat{\beta}(\theta_0)$ is a complete sufficient statistic for β, so that by Basu's theorem (Lehmann, 1983, p. 46) $\hat{\beta}(\theta_0)$ is distributed independently of

$$\hat{\beta}(\hat{\theta}) - \hat{\beta}(\theta_0)$$

since the latter has distribution not depending on β. Thus, we can write the covariance matrix of $\hat{\beta}(\hat{\theta})$ as

$$\text{Cov}\{N^{1/2}[\hat{\beta}(\hat{\theta}) - \beta)]\} = \sigma^2 S_G^{-1} + NE\{[\hat{\beta}(\hat{\theta}) - \hat{\beta}(\theta_0)][\hat{\beta}(\hat{\theta}) - \hat{\beta}(\theta_0)]^T\}$$

This is the key step, since it is the cross-product term that is most difficult to handle. The full details even now are complex, and are provided by Rothenberg (1984), but the major idea can be seen by making the expansion

$$N[\hat{\beta}(\hat{\theta}) - \hat{\beta}(\theta_0)] = G_N N^{1/2}(\hat{\theta} - \theta_0)$$

where G_N is a matrix with a limit G_∞ as $N \to \infty$. If $N^{1/2}(\hat{\theta} - \theta_0)$ converges to a normal random variable with mean zero and covariance Ω, then we have the expression

$$\text{Cov}\{N^{1/2}[\hat{\beta}(\hat{\theta}) - \beta]\} = S_G^{-1} + N^{-1}G_N \Omega G_N^T$$
$$+ \text{(smaller-order terms)}$$

Thus, for linear regression with normally distributed errors and the variance not depending on the mean, the ordering by efficiency of estimates of β is the same as the ordering by efficiency of estimate of θ.

7.3.4 Score tests for variance functions

The score test when there are nuisance parameters is defined in Cox and Hinkley (1974, p. 324). In this section we indicate how to derive the score test when testing for heteroscedasticity of a specific form with normally distributed data. The model is (2.1), and we wish to test the hypothesis that $\theta = \theta_0$. Write the logarithm of the likelihood as $L(\beta, \sigma, \theta)$, so that

$$L(\beta, \sigma, \theta) = -\sum_{i=1}^{N} \left(\log\left[\sigma g(\mu_i(\beta), z_i, \theta)\right] + \frac{[y_i - f(x_i, \beta)]^2}{2\sigma^2 g^2(\mu_i(\beta), z_i, \theta)} \right)$$

Further, let $\lambda = (\beta, \sigma)$. From section 7.2, the score test statistic is given by

$$Q_1(\hat{\beta}, \hat{\sigma}, \theta_0)^T \iota^{\theta\theta}(\hat{\beta}, \hat{\sigma}, \theta_0) Q_1(\hat{\beta}, \hat{\sigma}, \theta_0) \qquad (7.28)$$

where

$$Q_1(\beta, \sigma, \theta) = L_\theta(\beta, \theta, \sigma) - \iota_{\theta\lambda}(\beta, \sigma, \theta)\iota_{\lambda\lambda}^{-1}(\beta, \sigma, \theta)L_\lambda(\beta, \theta, \sigma)$$

Define S_G as in Theorem 2.1 and recall that $v(i, \beta, \theta) = \log\left[g(\mu_i(\beta), z_i, \theta)\right]$. The general form of the test statistic can be computed by noting the following easily proved identities

$$\iota_{\theta\theta} = 2\sum_{i=1}^{N} v_\theta(i, \beta, \theta)v_\theta(i, \beta, \theta)^T$$

$$\iota_{\theta\sigma} = (2/\sigma)\sum_{i=1}^{N} v_\theta(i, \beta, \theta)$$

$$\iota_{\theta\beta} = 2\sum_{i=1}^{N} v_\theta(i, \beta, \theta)v_\beta(i, \beta, \theta)^T$$

$$\iota_{\sigma\sigma} = 2N/\sigma^2$$

$$l_{\sigma\beta} = (2/\sigma) \sum_{i=1}^{N} v_{\beta}(i, \beta, \theta)$$

$$l_{\beta\beta} = (N/\sigma^2)S_G + 2 \sum_{i=1}^{N} v_{\beta}(i, \beta, \theta)v_{\beta}(i, \beta, \theta)^{\mathsf{T}}$$

As discussed in section 3.3, there are important cases where major simplifications arise. First, given β and θ, $\hat{\sigma}$ is computed by maximizing $L(\beta, \sigma, \theta)$, so that

$$L_{\sigma}(\beta, \hat{\sigma}, \theta) = 0$$

Thus, partitioning the matrix $l_{\lambda\lambda}^{-1}$ as

$$l_{\lambda\lambda}^{-1} = \begin{bmatrix} l_{\beta\beta} & l_{\beta\sigma} \\ l_{\beta\sigma}^{\mathsf{T}} & l_{\sigma\sigma} \end{bmatrix}^{-1} = \begin{bmatrix} l^{\beta\beta} & l^{\beta\sigma} \\ l^{\sigma\beta} & l^{\sigma\sigma} \end{bmatrix}$$

and suppressing the dependence on $\hat{\beta}$ and $\hat{\sigma}$, we see that

$$Q_1(\hat{\beta}, \hat{\sigma}, \theta_0) = L_{\theta} - (l_{\theta\beta}l^{\beta\beta} + l_{\theta\sigma}l^{\sigma\beta})L_{\beta} \qquad (7.29)$$

If $\hat{\beta}$ is the maximum-likelihood estimate of β given θ_0, then $L_{\beta}(\hat{\beta}, \hat{\sigma}, \theta_0) = 0$, so that

$$Q_1(\hat{\beta}, \hat{\sigma}, \theta_0) = L_{\theta}(\beta, \hat{\sigma}, \theta_0)$$
$$= \sum_{i=1}^{N} [e_i^2(\hat{\beta}, \theta_0) - 1][v_{\theta}(i, \hat{\beta}, \theta_0) - \bar{v}_{\theta}(\hat{\beta}, \theta_0)] \qquad (7.30)$$

see equation (3.23).

The first special case for which major simplification occurs is if the variance does not depend on the mean, for which we find that

$$l_{\theta\beta} = l_{\sigma\beta} = 0$$

and hence that

$$l^{\theta\theta} = \tfrac{1}{2} \left(\sum_{i=1}^{N} d(i, \beta, \theta)d(i, \beta, \theta)^{\mathsf{T}} \right)^{-1},$$

which is just (3.24). Thus, when the variance does not depend on the mean and the maximum-likelihood estimate is used, i.e., fully iterated generalized least squares, then the score test statistic is (3.22).

We next consider the case that the variances depend on the mean. If fully iterated generalized least squares is used, then we find that

$$L_{\beta}(\hat{\beta}, \hat{\sigma}, \theta_0) = \sum_{i=1}^{N} [e_i^2(\hat{\beta}, \theta_0) - 1]d_*(i, \hat{\beta}, \theta_0)$$

where
$$d_*(i, \beta, \theta) = v_\beta(i, \beta, \theta) - \bar{v}_\beta(\beta, \theta)$$

Thus, as suggested in section 3.3, if under the hypothesis $v_\beta = 0$, then the score test statistic is (3.22).

The final special case occurs when σ converges to zero as N converges to infinity. From (7.30), it suffices to show that

$$\lim_{\substack{N \to \infty \\ \sigma \to 0}} (\iota_{\theta\beta} \iota^{\beta\beta} + \iota_{\theta\sigma} \iota^{\sigma\beta}) = 0$$

which is immediate.

7.3.5 Characterizing restricted maximum likelihood

The restricted maximum-likelihood estimate maximizes in θ the marginal posterior distribution (3.15). We want to show that this estimate is equivalent to solving (3.12). Recall that

$$e_i(\beta, \theta, \sigma) = [y_i - f(x_i, \beta)]/[\sigma g(\mu_i(\beta), z_i, \theta)]$$
$$r_i(\beta, \theta, \sigma) = \sigma e_i(\beta, \theta, \sigma)$$

and

$$\hat{\sigma}_G^2(\theta) = (N - p)^{-1} \sum_{i=1}^{N} r_i^2(\hat{\beta}, \theta, \sigma)$$

Taking logarithms, restricted maximum likelihood maximizes in θ

$$-\sum_{i=1}^{N} \log[g(\mu_i(\hat{\beta}), z_i, \theta)] - \tfrac{1}{2}(N - p)\hat{\sigma}_G^2(\theta) - \tfrac{1}{2}\log\{\mathrm{Det}[S_G(\theta)]\}$$

$$(7.31)$$

Recall that
$$v(i, \beta, \theta) = \log[g(\mu_i(\beta), z_i, \theta)]$$

and that v_θ is the derivative of v with respect to θ. Then setting equal to zero the derivative with respect to θ of (7.31) yields the equation

$$\sum_{i=1}^{N} [e_i^2(\hat{\beta}, \theta, \sigma) - 1]v_\theta(i, \hat{\beta}, \theta) = \frac{1}{2}\frac{\partial}{\partial\theta}\log\{\mathrm{Det}[S_G(\theta)]\} \quad (7.32)$$

From Nel (1980),

$$\frac{\partial}{\partial\theta}\log\{\mathrm{Det}[S_G(\theta)]\} = \mathrm{trace}\left(S_G(\theta)^{-1}\frac{\partial}{\partial\theta}S_G(\theta)\right) \quad (7.33)$$

Let X_* be the $N \times p$ matrix with ith row the transpose of (3.11), and let $\Omega(\theta)$ be the $N \times N$ diagonal matrix with ith diagonal element $v(i, \hat{\beta}, \theta)$. Then

$$\frac{\partial}{\partial \theta} S_G(\theta) = -2N^{-1} X_*^T \Omega(\theta) X_*(\theta) \qquad (7.34)$$

Plugging (7.34) into (7.33), we find that

$$\frac{1}{2}\frac{\partial}{\partial \theta} \log\{\text{Det}[S_G(\theta)]\} = -N^{-1} \text{trace}[S_G(\theta)^{-1} X_*^T \Omega(\theta) X_*(\theta)]$$

$$= -\text{trace}\{\Omega(\theta) X_*(\theta)[N S_G(\theta)]^{-1} X_*(\theta)^T\}$$

However, $N S_G(\theta) = X_*(\theta)^T X_*(\theta)$, so that the right-hand side of (7.32) becomes

$$-\text{trace}[\Omega(\theta) H] = -\sum_{i=1}^{N} v_\theta(i, \hat{\beta}, \theta) h_{ii}$$

Substituting into (7.32) shows the equivalence with (3.12).

CHAPTER 8

Some open problems

In this chapter, we briefly list some problems that could be of interest for further research.

(1) *Estimation of the mean response after transformation.* There are a variety of point estimates, such as the smearing estimator or the parametric estimator, that require numerical integration. These have been described in section 4.4. Questions of robustness, confidence intervals, testing, etc., remain.

(2) *Using outside information to estimate transformation and variance parameters.* In fields such as radioimmunoassay (RIA), many small experiments are performed at more or less the same time. Trying to estimate transformation and variance parameters is problematic if one uses only the data in each (small) experiment. However, it is often reasonable to assume that the transformation or variance parameter stays fixed from experiment to experiment. Alternatively, one might think of the transformation or variance parameters as random variables that vary across experiments. In either framework, pooling information across experiments should yield much better estimates of transformation and variance parameters. When the transformation and/or variance parameters are random, then empirical Bayes estimation (Morris, 1983) could be used.

(3) *Nonparametric estimation of transformation and variance functions.* In the context of transformation to a nearly normal distribution, it would be useful to have a nonparametric device to construct a first approximate transformation, possibly for later use in building a parametric model. Methods such as ACE (Breiman and Friedman 1985), the method of sieves (Grenander, 1981; Geman and Hwang, 1982), etc. need to be investigated further in the context of transformations to stabilize variance and symmetrize the error distribution. In the variance-function context, we have already proposed kernel

regression methods (Carroll, 1982a), but questions about optimal choice of bandwidth are largely unexplored.

(4) *High-breakdown robust estimation.* It would be very useful to have high-efficiency, high-breakdown robust estimates of transformation and weighting parameters. The first step would be the extension of the least-median-of-squares estimator (Rousseeuw, 1984) to nonlinear regression. For a fixed value of the transformation or variance parameters this would provide a high-breakdown estimator of the regression parameters. Then a robust estimator of the transformation or variance parameters could be applied to the residuals.

(5) *Measurement error.* The effects of and correction for measurement error in the response and the predictors is largely unexplored in transformation and weighting models.

(6) *Extension of dispersion and transformation function estimation (including robustness) to the nonindependent observations.* There are two important examples where the observations will not be independent, nested design models and time series, and weighting and transformations should be very useful in both cases.

(7) *Transformation and weighting when the model is not known a priori.* One might think of doing transform both sides with different transformations. Also, if the mean function is not fairly well understood, it might be that some form of replication should be designed into experiments to get a good understanding of the variance function, absolute residuals being somewhat problematic.

(8) *Treatment of zero responses.* (This problem was suggested to us by Mats Rudemo.) In many situations where the response is positive, measurements can be made to only a fixed number of decimal places so that some measured response will be zero. This problem can sometimes be corrected by adding a constant to all responses (or perhaps to only the zero responses), but as shown in the bacterial clearance example of Chapters 5 and 6 this technique can induce high-leverage points and outliers. In the bacterial clearance example, the outliers were accommodated by robust estimation.

An alternative method suggested by Niels Keiding and communicated to us by Rudemo would be to model the zero responses as censored data. Explicitly, any zero response would be considered a response known only to lie between 0 and C, where C is he limit of measurement. The limit C might be known in some situations, but could be unknown in others.

(9) *Computation of the estimators.* Most of estimators in this book are found by solving nonlinear equations, a potentially difficult numerical problem. Finding reliable computational methods is a very important task. In a Monte Carlo study of several estimators for random-coefficients regression models, Dent and Hildreth (1977) found several well known numerical methods unreliable for calculation of the maximum-likelihood estimator.

We have had good success with the MAXLIK routine supplied with the GAUSS matrix language written for personal computers. This routine uses the Fisher method of scoring for calculation of the direction of the next step and uses 'squeezing' to find the optimal length of this step. Squeezing consists of comparing the loglikelihood at the full scoring-method step and at half the step. If the loglikelihood is greater at half a step than at the full step, then the step is halved and this process is repeated up to a fixed number of times (we used 10). Squeezing or some other method for preventing wild initial steps seems essential.

References

Abramowitz, L. and Singh, K. (1985) Edgeworth corrected pivotal statistics and the bootstrap. *Ann. Stat.* **13**, 116–32.

Amemiya, T. (1973) Regression analysis when the variance of the dependent variable is proportional to the square of its expectation. *J. Am. Stat. Assoc.* **68**, 928–34.

Amemiya, T. (1977) A note on a heteroscedastic model. *J. Econometrics* **6**, 365–70; Corrigenda. *J. Econometrics* **8**, 275.

Andrews, D.F. (1971) A note on the selection of data transformation. *Biometrika* **58**, 249–54.

Anscombe, F.J. (1961) Examination of residuals. *Proc. Fourth Berkeley Symp. on Mathematical Statistics and Probability* vol. I, pp. 1–36. University of California Press, Berkeley.

Atkinson, A.C. (1982) Regression diagnostics, transformations and constructed variables (with discussion). *J. R. Stat. Soc. B* **44**, 1–36.

Atkinson, A.C. (1985) *Plots, Transformations and Regression*. Clarendon Press, Oxford.

Atkinson, A.C. (1986a) Diagnostic tests for transformations. *Technometrics* **28**, 29–38.

Atkinson, A.C. (1986b) Masking unmasked. *Biometrika* **73**, 533–42.

Barker, T.B. (1986) Quality engineering by design: Taguchi's philosophy. *Qual. Prog.* December, 32–42.

Bar-Lev, S.K. and Ennis, P. (1986) Reproducibility and natural exponential families with power variance functions. *Ann. Stat.* **14**, 1507–22.

Bartlett, M.S. (1947) The use of transformations. *Biometrics* **3**, 39–52.

Bates, D.M., Hamilton, D.C. and Watts, D.G. (1982) Accounting for intrinsic nonlinearity in nonlinear regression parameter inference regions. *Ann. Stat.* **10**, 386–93.

Bates, D.M. and Watts, D.G. (1980) Relative curvature measures and nonlinearity. *J. R. Stat. Soc. B* **42**, 1–25.

Bates, D.M. and Watts, D.G. (1988) *Nonlinear Regression. Analysis and its Applications*. Wiley, New York.

Bates, D.M., Wolf, D.A. and Watts, D.G. (1985) Nonlinear least squares and first order kinetics. *Proc. Computer Science and Statistics: Seventeenth Symp. on the Interface* ed. D. Allen. North-Holland, New York.

Baxter, L.A., Coutts, S.M. and Ross, G.A.F. (1980) Applications of linear models in motor insurance. *Proc. 21st Int. Congr. Actuaries*, Zurich, pp. 11–29.

Beal, S.L. and Sheiner, L.B. (1985) Heteroscedastic nonlinear regression with pharmacokinetic type data. Preprint.

Begun, J.M., Hall, W.J., Huang, W.M. and Wellner, J.A. (1983) Information and asymptotic efficiency in parametric–nonparametric models. *Ann. Stat.* **11**, 432–52.

Belsley, D.A., Kuh, E. and Welsch, R.E. (1980) *Regression Diagnostics: Identifying Influential Data and Sources of Collinearity*. Wiley, New York.

Beran, R.J. (1986) Discussion of Wu's paper. *Ann. Stat.* **14**, 1295–8.

Beverton, R.J.H. and Holt, S.J. (1957) *On the Dynamics of Exploited Fish Populations*. Her Majesty's Stationery Office, London.

Bickel, P.J. (1978) Using residuals robustly I: tests for heteroscedasticity, nonlinearity. *Ann. Stat.* **6**, 266–91.

Bickel, P.J. (1982) On adaptive estimation. *Ann. Stat.* **10**, 647–71.

Bickel, P.J. and Doksum, K.A. (1981) An analysis of transformations revisited. *J. Am. Stat. Assoc.* **76**, 296–311.

Blom, G. (1958) *Statistical Estimates and Transformed Beta Variables*. Wiley, New York.

Box, G.E.P. (1984) Recent research in experimental design for quality improvement with applications to logistics. Technical report.

Box, G.E.P. (1987) Signal to noise ratios, performance criteria and transformation. *Technometrics* **29**.

Box, G.E.P. and Cox, D.R. (1964) An analysis of transformations. *J. R. Stat. Soc. B* **26**, 211–46.

Box, G.E.P. and Cox, D.R. (1982) An analysis of transformations revisited. *J. Am. Stat. Assoc.* **77**, 209–10.

Box, G.E.P. and Hill, W.J. (1974) Correcting inhomogeneity of variance with power transformation weighting. *Technometrics* **16**, 385–9.

Box, G.E.P. and Jenkins, G.M. (1976) *Time Series Analysis: Forecasting and Control*. Holden-Day, San Francisco.

Box, G.E.P. and Meyer, R.D. (1985a) An analysis for unreplicated fractional factorials. *Technometrics* **28**, 11–18.

Box, G.E.P. and Meyer, R.D. (1985b) Dispersion effects from fractional designs. *Technometrics* **28**, 19–28.

Breiman, L. and Friedman, J.H. (1985) Estimating optimal transformations for multiple regression and correlation. *J. Am. Stat. Assoc.* **80**, 580–614.

Breusch, T.S. and Pagan, A.R. (1979) A simple test for heteroscedasticity and random coefficient variation. *Econometrica* **47**, 1287–94.

Butt, W.R. (1984) *Practical Immunoassay*. Dekker, New York.

Carroll, R.J. (1980) A robust method for testing transformations to achieve approximate normality. *J. R. Stat. Soc. B* **42**, 71–8.

Carroll, R.J. (1982a) Adapting for heteroscedasticity in linear models. *Ann. Stat.* **10**, 1224–33.

Carroll, R.J. (1982b) Power transformations when the choice of power is restricted to a finite set. *J. Am. Stat. Assoc.* **77**, 908–15.

Carroll, R.J. (1982c) Two examples of transformations where there are possible outliers. *Appl. Stat.* **31**, 149–52.

Carroll, R.J. (1983) Tests for regression parameters in power transformation models. *Scand. J. Stat.* **9**, 217–22.

Carroll, R.J. and Cline, D.B.H. (1988) An asymptotic theory for weighted least squares with weights estimated by replication. *Biometrika* **75**.

Carroll, R.J., Davidian, M. and Smith, W.C. (1986a) Variance functions and the minimum detectable concentration in assays. University of North Carolina Technical Report 1701.

Carroll, R.J. and Ruppert, D. (1981a) On prediction and the power transformation family. *Biometrika* **68**, 609–15.

Carroll, R.J. and Ruppert, D. (1981b) On robust tests for heteroscedasticity. *Ann. Stat.* **9**, 206–10.

Carroll, R.J. and Ruppert, D. (1982a) A comparison between maximum likelihood and generalized least squares in a heteroscedastic linear model. *J. Am. Stat. Assoc.* **77**, 878–82.

Carroll, R.J. and Ruppert, D. (1982b) Robust estimation in heteroscedastic linear models. *Ann. Stat.* **10**, 429–41.

Carroll, R.J. and Ruppert, D. (1984a) Power transformations when fitting theoretical models to data. *J. Am. Stat. Assoc.* **79**, 321–8.

Carroll, R.J. and Ruppert, D. (1984b) Discussion of Hinkley and Runger's paper 'The analysis of transformed data'. *J. Am. Stat. Assoc.* **79**, 312–13.

Carroll, R.J. and Ruppert, D. (1985) Transformations: a robust analysis. *Technometrics* 27, 1–12.

Carroll, R.J. and Ruppert, D. (1986) Discussion of Wu's paper. *Ann. Stat.* 14, 1298–1301.

Carroll, R.J. and Ruppert, D. (1987) Diagnostics and robust estimation when transforming the regression model and the response. *Technometrics* 29, 287–99.

Carroll, R.J., Ruppert, D. and Stefanski, L.A. (1986b) Adapting for heteroscedasticity in regression models. University of North Carolina Technical Report 1702.

Carroll, R.J., Sacks, J. and Spiegelman, C.H. (1987) A quick and easy multiple use calibration curve procedure. *Technometrics* 29.

Carroll, R.J., Wu, C.F.J. and Ruppert, D. (1987) The effect of estimating weights in generalized least squares. Preprint.

Cochran, W.G. (1937) Problems arising in the analysis of a series of similar experiments. *J. R. Stat. Soc.* Suppl. 4, 102–18.

Cohen, M.L., Dalal, S.R. and Tukey, J.W. (1984) Robust smoothly heterogeneous variance regression. Preprint.

Cook, R.D. (1977) Detection of influential observations in linear regression. *Technometrics* 19, 15–18.

Cook, R.D. (1986) Assessment of local influence. *J. R. Stat. Soc. B* 48, 133–69.

Cook, R.D. and Tsai, C.L. (1985) Residuals in nonlinear regression. *Biometrika* 72, 23–9.

Cook, R.D. and Weisberg, S. (1982) *Residuals and Influence in Regression*. Chapman and Hall, New York and London.

Cook, R.D. and Weisberg, S. (1983) Diagnostic for heteroscedasticity in regression. *Biometrika* 70, 1–10.

Cox, D.R. and Hinkley, D.V. (1974) *Theoretical Statistics*. Chapman and Hall, New York.

Cressie, N.A.C. and Keightley, D.D. (1979) The underlying structure of the direct linear plot with applications to the analysis of hormone–receptor interactions. *J. Steroid Biochem.* 11, 1173–80.

Cressie, N.A.C. and Keightley, D.D. (1981) Analyzing data from hormone–receptor assays. *Biometrics* 37, 325–40.

Currie, D.J. (1982) Estimating Michaelis–Menten parameters: bias, variance and experimental design. *Biometrics* 38, 907–19.

Davidian, M. (1986) Variance function estimation in heteroscedastic regression models. Ph.D. Dissertation, University of North Carolina at Chapel Hill.

Davidian, M. and Carroll, R.J. (1987) Variance function estimation. *J. Am. Stat. Assoc.* **82**, 1079–91.

Davidian, M. and Carroll, R.J. (1988) An analysis of extended quasi-likelihood. *J. R. Stat. Soc. B* **50**.

Dent, W.T. and Hildreth, C. (1977) Maximum likelihood estimation in random coefficient models. *J. Am. Stat. Assoc.* **72**, 69–72.

Doksum, K.A. and Wong, C.W. (1983) Statistical tests based on transformed data. *J. Am. Stat. Assoc.* **78**, 411–16.

Draper, N. and Smith, H. (1981) *Applied Regression Analysis*, 2nd edn. Wiley, New York.

Duan, N. (1983) Smearing estimate: a nonparametric retransformation method. *J. Am. Stat. Assoc.* **78**, 605–10.

Eaton, M.L. (1985) The Gauss–Markov theorem in multivariate analysis. *Multivariate Analysis* vol. VI, ed. P.R. Krishnaiah. Elsevier, Amsterdam.

Efron, B. (1979) Bootstrap methods: another look at the jackknife. *Ann. Stat.* **7**, 1–26.

Efron, B. (1982) *The Jackknife, the Bootstrap, and Other Resampling Plans*. CBMS–NSF Monograph 39. SIAM, Philadelphia.

Efron, B. (1986) Double exponential families and their use in generalized linear regression. *J. Am. Stat. Assoc.* **81**, 709–21.

Efron, B. (1987) Bootstrap confidence intervals and bootstrap approximations (with discussion). *J. Am. Stat. Assoc.* **82**, 171–200.

Efron, B. and Hinkley, D.V. (1978) Assessing the accuracy of the maximum likelihood estimator: observed versus expected Fisher information. *Biometrika* **65**, 457–81.

Efron, B. and Tibshirani, R. (1986) Bootstrap methods for standard errors, confidence intervals and other measures of statistical accuracy. *Stat. Sci.* **1**, 54–77.

Egy, D. and Lahiri, K. (1979) On maximum likelihood estimation of functional forms and heteroscedasticity. *Economics Lett.* **2**, 155–60.

Finney, D.J. (1976) Radioligand assay. *Biometrics* **32**, 721–40.

Firth, D. (1987) On the efficiency of quasi-likelihood estimation. *Biometrika* **74**, 233–46.

Freedman, D.A. and Peters, S.C. (1984) Bootstrapping a regression equation: some empirical results. *J. Am. Stat. Assoc.* **79**, 97–104.

Fuller, W.A. (1987) *Measurement Error Models*. Wiley, New York.

Fuller, W.A. and Rao, J.N.K. (1978) Estimation for a linear regression model with unknown diagonal covariance matrix. *Ann. Stat.* **6**, 1149–58.

Garden, J.S., Mitchell, D.G. and Mills, W.N. (1980) Nonconstant variance regression techniques for calibration curve based analysis. *Anal. Chem.* **52**, 2310–15.

Geman, S. and Hwang, C.R. (1982) Nonparametric maximum likelihood by the method of sieves. *Ann. Stat.* **10**, 401–4.

Giltinan, D.M., Carroll, R.J. and Ruppert, D. (1986) Some new methods for weighted regression when there are possible outliers. *Technometrics* **28**, 219–30.

Glejser, H. (1969) A new test for heteroscedasticity. *J. Am. Stat. Assoc.* **64**, 316–23.

Goldberger, A.S. (1964) *Econometric Theory.* Wiley, New York.

Goldfeld, S.M. and Quandt, R.E. (1972) *Nonlinear Methods in Econometrics.* North-Holland, Amsterdam.

Green, P.J. (1985) Linear models for field trials, smoothing and cross-validation. *Biometrika* **72**, 527–38.

Grenander, U. (1981) *Abstract Inference.* Wiley New York.

Haerdle, W. (1984) Robust regression function estimation. *J. Multivariate Anal.* **14**, 169–80.

Haerdle, W. and Gasser, T. (1984) Robust nonparametric function fitting. *J. R. Stat. Soc.* B **46**, 42–51.

Haerdle, W., Hall, P. and Marron, J.S. (1987) How far are automatically chosen regression smoothing parameters from their optimum? *J. Am. Stat. Assoc.* **82**.

Haerdle, W. and Tsybakov, A.B. (1987) Robust nonparametric regression with simultaneous scale curve estimation. *Ann. Stat.* **15**.

Hall, P. (1988) Theoretical comparison of bootstrap confidence intervals. *Ann. Stat.* **16**.

Hall, P. and Marron, J.S. (1987) On the amount of noise inherent in bandwidth selection for a kernel density estimator. *Ann. Stat.* **15**, 163–81.

Hampel, F.R. (1968) Contributions to the theory of robust estimation. Ph.D. Thesis, University of California at Berkeley.

Hampel, F.R. (1971) A general qualitative definition of robustness. *Ann. Math. Stat.* **42**, 1887–96.

Hampel, F.R. (1974) The influence curve and its role in robust estimation. *J. Am. Stat. Assoc* **62**, 1179–86.

Hampel, F.R. (1985) The breakdown point of the mean combined with some rejection rules. *Technometrics* **27**, 95–107.

Hampel, F.R., Ronchetti, E.M., Rousseeuw, P.J. and Stahel, W.A.

(1986) *Robust Statistics: The Approach Based on Influence Functions.* Wiley, New York.

Harrison, M.J. and McCabe, B.P.M. (1979) A test for heteroscedasticity based on least squares residuals. *J. Am. Stat. Assoc.* **74**, 494–500.

Harvey, A.C. (1976) Estimating regression models with multiplicative heteroscedasticity. *Econometrics* **44**, 461–5.

Harville, D. (1977) Maximum likelihood approaches to variance component estimation and to related problems. *J. Am. Stat. Assoc.* **72**, 320–40.

Hernandez, F. and Johnson, R.A. (1980) The large sample behavior of transformations to normality. *J. Am. Stat. Assoc.* **75**, 855–61.

Hildreth, C. and Houck, J.P. (1968) Some estimates for a linear model with random coefficients. *J. Am. Stat. Assoc.* **63**, 584–95.

Hinkley, D.V. (1975) On power transformations to symmetry. *Biometrika* **62**, 101–11.

Hinkley, D.V. and Runger, G. (1984) Analysis of transformed data. *J. Am. Stat. Assoc.* **79**, 302–8.

Hoaglin, D.C., Mosteller, F. and Tukey, J.W. (1983) *Understanding Robust and Exploratory Data Analysis.* Wiley, New York.

Hoaglin, D.C., Mosteller, F. and Tukey, J.W. (1985) *Exploring Data Tables, Trends and Shapes.* Wiley, New York.

Hopwood, W.S., McKeown, J.C. and Newbold, P. (1984) Time series forecasting models involving power transformations. *J. Forecasting* **3**, 57–61.

Huber, P.J. (1964) Robust estimation of a location parameter. *Ann. Math. Stat.* **35**, 73–101.

Huber, P.J. (1967) The behavior of maximum likelihood estimates under nonstandard conditions. *Proc. Fifth Berkeley Symp. on Mathematical Statistics and Probability*, vol. 1, pp. 221–33.

Huber, P.J. (1981) *Robust Statistics.* Wiley, New York.

Jacquez, J.A., Mather, F.J. and Crawford, C.R. (1968) Linear regression with non-constant, unknown error variances: sampling experiments with least squares and maximum likelihood estimators. *Biometrics* **24**, 607–26.

Jacquez, J.A. and Norusis, M. (1973) Sampling experiments on the estimation of parameters in heteroscedastic linear regression. *Biometrics* **29**, 771–80.

Jobson, J.D. and Fuller, W.A. (1980) Least squares estimation when the covariance matrix and parameter vector are functionally related. *J. Am. Stat. Assoc.* **75**, 176–81.

Judge, G.G., Griffiths, W.E., Hill, R.C., Lutkepohl, H. and Lee, T.C. (1985) *The Theory and Practice of Econometrics* 2nd edn. Wiley, New York.

Just, R.E. and Pope, R.D. (1978) Stochastic specification of production functions and economic implications. *J. Econometrics* **7**, 67–86.

Kirkwood, T.B.L. (1977) Predicting the stability of biological standards and products. *Biometrics* **33**, 736–42.

Kirkwood, T.B.L. (1984) Design and analysis of accelerated degradation tests for the stability of biological standards III: principles of design. *J. Biol. Stand.* **12**, 215–24.

Koenker, R. and Bassett, G. (1981) Robust tests for heteroscedasticity based on regression quantiles. *Econometrica* **50**, 43–61.

Kotz, S. and Johnson, N.L. (1985) *Encyclopedia of Statistical Sciences* vol. 5. Wiley, New York.

Krasker, W.S. and Welsch, R.E. (1982) Efficient bounded influence regression estimation. *J. Am. Stat. Assoc.* **77**, 595–604.

Leech, D. (1975) Testing the error specification in nonlinear regression. *Econometrica* **43**, 719–25.

Lehmann, E.L. (1983) *Theory of Point Estimation.* Wiley, New York.

Leon, R.V., Shoemaker, A.C. and Kackar, R.N. (1987) Performance measures independent of adjustment: an explanation and extension of Taguchi's signal to noise ratios. *Technometrics* **29**, 253–82.

Leurgans, S. (1980) Evaluating laboratory measurement techniques. *Biostatistics Casebook* eds R.G. Miller Jr, B. Efron, B.W. Brown Jr and L.E. Moses. Wiley, New York.

Lieberman, G.J., Miller, R.G. and Hamilton, M.A. (1967) Unlimited simultaneous discrimination intervals in regression. *Biometrika* **54**, 133–45.

Maronna, R.A. (1976) Robust estimation of multivariate location and scatter. *Ann. Stat.* **4**, 51–67.

Marron, J.S. (1986) Will the art of smoothing ever become a science? *Function Estimates* ed. J.S. Marron. Am. Math. Soc. Contemp. Math. Ser.

Marron, J.S. (1987) A comparison of cross-validation techniques in density estimation. *Ann. Stat.* **15**, 152–63.

Matloff, N., Rose, R. and Tai, R. (1984) A comparison of two methods for estimating optimal weights in regression analysis. *J. Stat. Comput. Simulation* **19**, 265–74.

McCullagh, P. (1983) Quasi-likelihood functions. *Ann. Stat.* **11**, 59–67.

McCullagh, P. and Nelder, J.A. (1983) *Generalized Linear Models*. Chapman and Hall, New York.

McCullagh, P. and Pregibon, D. (1987) K-statistics in regression. *Ann. Stat.* **16**, 202–19.

Miller, R.G., Jr (1981) *Simultaneous Statistical Inference* 2nd Edn. Springer, New York.

Morris, C.N. (1983) Parametric empirical Bayes inference: theory and applications (with discussion). *J. Am. Stat. Assoc.* **78**, 47–65.

Morton, R. (1987a) A generalized linear model with nested strata of extra-Poisson variation. *Biometrika* **74**, 247–58.

Morton, R. (1987b) Analysis of generalized linear models with nested strata of variation. Preprint.

Muller, H.G. and Stadtmuller, U. (1986) Estimation of heteroscedasticity in regression analysis. Preprint.

Myers, L.E., Sexton, N.H., Southerland, L.I. and Wolff, T.J. (1981) Regression analysis of Ames test data. *Environ. Mutagenesis* **3**, 575–86.

Nel, D.G. (1980) On matrix differentiation in statistics. *S. Afr. Stat. J.* **14**, 87–101.

Nelder, J.A. and Pregibon, D. (1987) An extended quasi-likelihood function. *Biometrika* **74**, 221–32.

Nelson, P.R. (1983) Stability prediction using the Arrhenius model. *Comput. Programs Biomed.* **16**, 55–60.

Neyman, J. and Scott, E. (1948) Consistent estimates based on partially consistent observations. *Econometrica* **16**, 1–32.

Oja, H. (1981) On location, scale, skewness and kurtosis of univariate distributions. *Scand. J. Stat.* **8**, 154–68.

Oppenheimer, L., Capizzi, T.P., Weppelman, R.M. and Mehto, H. (1983) Determining the lowest limit of reliable assay measurement. *Anal. Chem.* **55**, 638–43.

Patefield, W.M. (1977) On the maximized likelihood function. *Sankhya B* **39**, 92–6.

Patterson, H.D. and Thompson, R. (1971) Recovery of inter-block information when block sizes are unequal. *Biometrika* **58**, 545–54.

Pierce, D.A. and Schafer, D.W. (1986) Residuals in generalized linear models. *J. Am. Stat. Assoc.* **81**, 977–86.

Pritchard, D.J., Downie, J. and Bacon, D.W. (1977) Further consideration of heteroscedasticity in fitting kinetic models. *Technometrics* **19**, 227–36.

Raab, G.M. (1981a) Estimation of a variance function, with application to radioimmunoassay. *Appl. Stat.* **30**, 32–40.

Raab, G.M. (1981b) Letter on 'Robust calibration and radioim-munoassay'. *Biometrics* **37**, 839–41.

Rao, C.R. (1973) *Linear Statistical Inference and its Applications* 2nd edn. Wiley, New York.

Rao, J.N.K. and Subrahmanian, K. (1971) Combining independent estimators in linear regression with unequal variances. *Biometrics* **27**, 971–90.

Ratkowsky, D.A. (1983) *Nonlinear Regression Modeling*. Dekker, New York.

Reish, R.L., Deriso, R.B., Ruppert, D. and Carroll, R.J. (1985) An investigation of the population dynamics of Atlantic menhaden (*Brevoortia tyrannus*). *Can. J. Fish. Aquatic Sci.* **42**, 147–57.

Ricker, W.E. (1954) Stock and recruitment. *J. Fish. Res. Board Can.* **11**, 559–623.

Ricker, W.E. and Smith, H.D. (1975) A revised interpretation of the history of the Skeena River sockeye salmon. *J. Fish. Res. Board Can.* **32**, 1369–81.

Robinson, P.M. (1987) Asymptotically efficient estimation in the presence of heteroscedasticity of unknown form. *Econometrica* **55**, 875–92.

Rodbard, D. (1978) Statistical estimation of the minimum detectable concentration ('sensitivity') for radioligand assays. *Anal. Biochem.* **90**, 1–12.

Rodbard, D. and Frazier, G.R. (1975) Statistical analysis of radio-ligand assay data. *Meth. Enzymol.* **37**, 3–22.

Rosenblatt, J.R. and Spiegelman, C.H. (1981) Discussion of the paper by Hunter and Lamboy. *Technometrics* **23**, 329–33.

Rothenberg, T.J. (1984) Approximate normality of generalized least squares estimates. *Econometrica* **52**, 811–25.

Rousseeuw, P.J. (1984) Least median of squares regression. *J. Am. Stat. Assoc.* **79**, 871–80.

Rousseeuw, P.J. and Yohai, V.J. (1984) Robust regression by means of S-estimators. *Robust and Nonlinear Time Series Analysis* eds J. Franke, W. Haerdle and R.D. Martin. Springer, Berlin.

Ruppert, D. (1985) On the bounded influence regression estimator of Krasker and Welsch. *J. Am. Stat. Assoc.* **80**, 205–8.

Ruppert, D. (1987) What is kurtosis? *Am. Statistician* **41**, 1–5.

Ruppert, D. and Carroll, R.J. (1980) Trimmed least squares estim-ation in the linear model. *J. Am. Stat. Assoc.* **77**, 828–38.

Ruppert, D. and Carroll, R.J. (1985) Data transformations in regression analysis with applications to stock recruitment relationships. *Resource Management* ed. M. Mangel. Lecture Notes in Biomathematics 61. Springer, New York.

Ruppert, D., Jakab, G.J., Sylwester, D.L. and Green, G.M. (1975) Sources of variance in the measurement of intrapulmonary killing of bacteria. *J. Lab. Clin. Med.* **87**, 544–58.

Ruppert, D., Reish, R.L., Deriso, R.B. and Carroll, R.J. (1985) A stochastic model for managing the Atlantic menhaden fishery and assessing managerial risks. *Can. J. Fish. Aquatic Sci.* **42**, 1371–9.

Sadler, W.A. and Smith, M.H. (1985) Estimation of the response–error relationship in immunoassay. *Clin. Chem.* **31** (11), 1802–5.

Scheffe, H. (1973) A statistical theory of calibration. *Ann. Stat.* **1**, 1–37.

Schwartz, L.M. (1979) Calibration curves with nonuniform variance. *Anal. Chem.* **51**, 723–9.

Seber, G.A.F. (1977) *Linear Regression Analysis*. Wiley, New York.

Silverman, B.W. (1985) Some aspects of the spline smoothing approach to nonparametric regression curve fitting (with discussion). *J. R. Stat. Soc. B* **46**, 1–52.

Snee, R.D. (1986) An alternative approach to fitting models when reexpression of the response is useful. *J. Qual. Technol.* **18**, 211–25.

Snee R.D. and Irr, J.D. (1981) Design of a statistical method for the analysis of mutagenesis at the HGPRT locus of cultured chinese hamster ovary cells. *Mutation Res.* **85**, 77–93.

Stefanski, L.A., Carroll, R.J. and Ruppert, D. (1986) Optimally bounded score functions for generalized linear models with applications to logistic regression. *Biometrika* **73**, 413–24.

Stirling, W.D. (1985) Heteroscedastic models and an application to block designs. *Appl. Stat.* **34**, 33–41.

Taguchi, G. and Wu, Y. (1980) *Introduction to Off-line Quality Control*. Central Japan Quality Control Association, Nagoya.

Taylor, J.M.G. (1985) Power transformations to symmetry. *Biometrika* **72**, 145–52.

Taylor, J.M.G. (1986) The retransformed mean after a fitted power transformation. *J. Am. Stat. Assoc.* **81**, 114–19.

Toyooka, Y. (1982) Second order expansion of mean squared error matrix of generalized least squares estimators with estimated parameters. *Biometrika* **69**, 269–73.

Tydeman, M.S. and Kirkwood, T.B.L. (1984) Design and analysis of accelerated degradation tests for the stability of biological standards I: properties of maximum likelihood estimators. *J. Biol. Stand.* **12**, 195–206.

van Zwet, W.R. (1964) *Convex Transformations of Random Variables.* Mathematisch Centrum, Amsterdam.

Wald, A. (1943) Tests of statistical hypotheses concerning several parameters when the number of observations is large. *Trans. Am. Math. Soc.* **54**, 426–82.

Watters, R.L., Carroll, R.J. and Spiegelman, C.H. (1987) Error modeling and confidence interval estimation for inductively coupled plasma calibration curves. *Anal. Chem.* **59**, 1639–43.

Wedderburn, R.W.M. (1974) Quasi-likelihood functions, generalized linear models and the Gauss–Newton method. *Biometrika* **61**, 439–47.

Weisberg, S. (1985) *Applied Linear Regression.* Wiley, New York.

Welsh, A. (1987) One step *L*-estimators in the linear model. *Ann. Stat.* **15**, 626–41.

Williams, E. (1959) *Regression Analysis.* Wiley, New York.

Williams, J.S. (1975) Lower bounds on convergence rates of weighted least squares to best linear unbiased estimators. *A Survey of Statistical Design and Linear Models* ed. J.N. Srivastava. North-Holland, Amsterdam.

Wu, C.F.J. (1986) Jackknife, bootstrap and other resampling plans in regression analysis (with discussion). *Ann. Stat.* **14**, 1261–350.

Yates, F. and Cochran, W.G. (1938) The analysis of groups of experiments. *J. Agric. Sci.* **28**, 556–80; reprinted (1970) in *Experimental Design: Selected Papers of Frank Yates.* Griffin, London.

Yohai, V. (1987) High breakdown point and high efficiency robust estimation for regression. *Ann. Stat.* **15**, 642–56.

Author Index

Subject Index